VOLUME SIX

ADVANCES IN
STEM CELLS AND THEIR NICHES

Recapitulating the Stem Cell Niche Ex Vivo

VOLUME SIX

Advances in
STEM CELLS AND THEIR NICHES

Recapitulating the Stem Cell Niche Ex Vivo

Edited by

SUSIE NILSSON

Biomedical Manufacturing Commonwealth Scientific and Industrial Research Organisation (CSIRO), Melbourne, Australia;
Australian Regenerative Medicine Institute, Monash University, Melbourne, Australia

ELSEVIER

ACADEMIC PRESS
An imprint of Elsevier

Academic Press is an imprint of Elsevier
50 Hampshire Street, 5th Floor, Cambridge, MA 02139, United States
525 B Street, Suite 1650, San Diego, CA 92101, United States
The Boulevard, Langford Lane, Kidlington, Oxford OX5 1GB, United Kingdom
125 London Wall, London EC2Y 5AS, United Kingdom

First edition 2022

Notices
Knowledge and best practice in this field are constantly changing. As new research and experience
broaden our understanding, changes in research methods, professional practices, or medical
treatment may become necessary.

Practitioners and researchers must always rely on their own experience and knowledge in evaluating
and using any information, methods, compounds, or experiments described herein. In using such
information or methods they should be mindful of their own safety and the safety of others, including
parties for whom they have a professional responsibility.

To the fullest extent of the law, neither the Publisher nor the authors, contributors, or editors, assume
any liability for any injury and/or damage to persons or property as a matter of products liability,
negligence or otherwise, or from any use or operation of any methods, products, instructions, or ideas
contained in the material herein.

ISBN: 978-0-323-91091-0
ISSN: 2468-5097

For information on all Academic Press publications
visit our website at https://www.elsevier.com/books-and-journals

Publisher: Zoe Kruze
Editorial Project Manager: Naiza Ermin Mendoza
Production Project Manager: James Selvam
Cover Designer: Miles Hitchen

Typeset by STRAIVE, India

Working together
to grow libraries in
developing countries

www.elsevier.com • www.bookaid.org

Contents

Contributors   *vii*

1. Recapitulating the liver niche *in vitro*   **1**

Kiryu K. Yap and Geraldine M. Mitchell

1. Introduction   2
2. The liver microenvironment   2
3. Cell sources to recreate the liver niche   7
4. Extracellular matrix and scaffolds   10
5. Bio-engineered platforms to recapitulate the liver niche   17
6. *In vitro* applications of bio-engineered liver platforms   29
7. Future directions   33
8. Conclusion   35
Acknowledgments   36
Conflict of interest   36
References   36

2. Organoid systems for recapitulating the intestinal stem cell niche and modeling disease *in vitro*   **57**

Hui Yi Grace Lim, Lana Kostic, and Nick Barker

1. Introduction   57
2. Intestinal stem cells and their *in vivo* niche   60
3. Modeling the ECM using synthetic matrices   63
4. Development of "mini-gut" organoid models   67
5. Organoids as a platform for modeling intestinal regeneration, disorders, and cancer   72
6. Conclusions   83
Acknowledgments   84
References   84

3. Reconstructing the lung stem cell niche *in vitro*   **97**

Dayanand Swami, Jyotirmoi Aich, Bharti Bisht, and Manash K. Paul

1. Introduction   99
2. Lung development   100
3. Lung stem cell diversity   102
4. Lung regeneration   107

5. Lung stem cell niche   109
6. Recapitulating stem cell niche   116
7. Culture systems and morphogens to study lung stem cell niche   120
8. Artificial scaffold and lung stem/progenitor cell niche   124
9. Conclusion and future direction   131
Acknowledgments   132
Competing interests   132
References   133

4. Engineering mammary tissue microenvironments *in vitro*   **145**
Julien Clegg, Maria Koch, Akhilandeshwari Ravichandran,
Dietmar W. Hutmacher, and Laura J. Bray

1. Anatomy of the normal mammary microenvironment   145
2. Development, progression and clinical management of BC   148
3. 3D tumor modeling   156
4. Current limitations and considerations of 3D models for BC   169
5. Future prospective of 3D BC modeling systems   170
References   172

5. Recapitulating human skeletal muscle in vitro   **179**
Anna Urciuolo, Maria Easler, and Nicola Elvassore

1. Lessons from skeletal muscle anatomy and physiology   183
2. Lesson from skeletal muscle development   185
3. Lesson from skeletal muscle regeneration   187
4. Emerging approaches for human skeletal muscle in vitro models   191
5. Conclusions and future perspectives   198
Acknowledgments   200
References   200

Contributors

Jyotirmoi Aich
School of Biotechnology and Bioinformatics, D. Y. Patil Deemed to Be University, CBD Belapur, Navi Mumbai, Maharashtra, India

Nick Barker
A★STAR Institute of Molecular and Cell Biology; Department of Physiology, Yong Loo Lin School of Medicine, National University of Singapore, Singapore, Singapore; Kanazawa University, Kanazawa, Japan

Bharti Bisht
Division of Thoracic Surgery, David Geffen School of Medicine, University of California Los Angeles, Los Angeles, CA, United States

Laura J. Bray
School of Mechanical, Medical and Process Engineering, Queensland University of Technology (QUT); ARC Training Centre for Cell and Tissue Engineering Technologies, Queensland University of Technology (QUT), Brisbane, QLD, Australia

Julien Clegg
School of Mechanical, Medical and Process Engineering, Queensland University of Technology (QUT); ARC Training Centre for Cell and Tissue Engineering Technologies, Queensland University of Technology (QUT), Brisbane, QLD, Australia

Maria Easler
Institute of Pediatric Research (IRP), Fondazione Città della Speranza, Padova, Italy

Nicola Elvassore
Department of Industrial Engineering, University of Padua; Veneto Institute of Molecular Medicine, Padova, Italy; University College London Great Ormond Street Institute of Child Health, London, United Kingdom

Dietmar W. Hutmacher
School of Mechanical, Medical and Process Engineering, Queensland University of Technology (QUT); ARC Training Centre for Cell and Tissue Engineering Technologies, Queensland University of Technology (QUT); School of Biomedical Sciences, Queensland University of Technology; Max Planck Queensland Center for Materials Science of Extracellular Matrices, Brisbane, QLD, Australia

Maria Koch
School of Mechanical, Medical and Process Engineering, Queensland University of Technology (QUT), Brisbane, QLD, Australia

Lana Kostic
A★STAR Institute of Molecular and Cell Biology, Singapore, Singapore

Hui Yi Grace Lim
A★STAR Institute of Molecular and Cell Biology, Singapore, Singapore

Geraldine M. Mitchell
O'Brien Institute, Department of St Vincent's Institute of Medical Research; University of
Melbourne, Department of Surgery, St Vincent's Hospital Melbourne; Australian Catholic
University, Fitzroy, Vic, Australia

Manash K. Paul
Division of Pulmonary and Critical Care Medicine, David Geffen School of Medicine,
University of California Los Angeles (UCLA), Los Angeles, CA, United States

Akhilandeshwari Ravichandran
School of Mechanical, Medical and Process Engineering, Queensland University of
Technology (QUT); ARC Training Centre for Cell and Tissue Engineering Technologies,
Queensland University of Technology (QUT), Brisbane, QLD, Australia

Dayanand Swami
School of Biotechnology and Bioinformatics, D. Y. Patil Deemed to Be University, CBD
Belapur, Navi Mumbai, Maharashtra, India

Anna Urciuolo
Institute of Pediatric Research (IRP), Fondazione Città della Speranza; Department of
Molecular Medicine, University of Padua, Padova, Italy

Kiryu K. Yap
O'Brien Institute, Department of St Vincent's Institute of Medical Research; University of
Melbourne, Department of Surgery, St Vincent's Hospital Melbourne, Fitzroy, Vic, Australia

CHAPTER ONE

Recapitulating the liver niche *in vitro*

Kiryu K. Yap[a,b] and Geraldine M. Mitchell[a,b,c],*
[a]O'Brien Institute, Department of St Vincent's Institute of Medical Research, Fitzroy, Vic, Australia
[b]University of Melbourne, Department of Surgery, St Vincent's Hospital Melbourne, Fitzroy, Vic, Australia
[c]Australian Catholic University, Fitzroy, Vic, Australia
*Corresponding author: e-mail address: gmitchell@svi.edu.au

Contents

1. Introduction   2
2. The liver microenvironment   2
    2.1 Micro-structural organization of the liver lobule   3
    2.2 Liver zonation   4
    2.3 Extracellular matrix   5
3. Cell sources to recreate the liver niche   7
    3.1 Primary cells   7
    3.2 Adult stem/progenitor cells   8
    3.3 Pluripotent stem cells   8
    3.4 Cell lines   9
4. Extracellular matrix and scaffolds   10
    4.1 Cell culture surface coatings   12
    4.2 Decellularized ECM matrices   12
    4.3 Hydrogels   15
    4.4 Fabricated scaffolds   16
5. Bio-engineered platforms to recapitulate the liver niche   17
    5.1 Two-dimensional (2D) platforms   17
    5.2 Three-dimensional (3D) platforms   20
    5.3 Organoids   22
    5.4 Repopulated decellularized liver ECM scaffolds   25
    5.5 Perfusion bioreactors and microfluidic systems   26
6. *In vitro* applications of bio-engineered liver platforms   29
    6.1 Developmental biology and normal physiology   29
    6.2 Disease modeling and target discovery   30
    6.3 Drug testing   32
7. Future directions   33
    7.1 Additional cells and structures   33
    7.2 Physiological cues   34
    7.3 Designer matrices   34
    7.4 Liver zonation   35
8. Conclusion   35

Advances in Stem Cells and their Niches, Volume 6
ISSN 2468-5097
https://doi.org/10.1016/bs.asn.2021.10.002

1

Acknowledgments   36
Conflict of interest   36
References   36

1. Introduction

The liver is the largest internal organ in the body, weighing 2% of the body or approximately 1.5 kg in a 70 kg adult human (Chan et al., 2006; Molina & DiMaio, 2012), and containing approximately 300 billion cells (Bianconi et al., 2013). It is a remarkably complex organ with over 500 different functions, including the production of major blood proteins (albumin, transferrin, globulins), coagulation factors, xenobiotic detoxification, nutrient metabolism, energy homeostasis, immune surveillance, and bile production (Ben-Moshe & Itzkovitz, 2019; Boyer & Soroka, 2021; Heymann & Tacke, 2016; Trefts, Gannon, & Wasserman, 2017). These functions arise from interactions between the liver's cells and its microenvironment.

To recreate this in an *in vitro* setting requires careful consideration of the cellular constituents, extracellular matrix, and mechano–physical factors such as environmental stiffness, fluid perfusion, and shear stress. Such *in vitro* models promise to offer novel insights into liver biology, with applications in studying developmental biology and disease models, and in drug development. Increasingly, these models are being constructed using human cells, providing clinically relevant models with practical applications. In this chapter, we will discuss the major concepts and techniques involved in reconstructing the liver niche, with a focus on human bioengineered liver platforms.

2. The liver microenvironment

In humans, the liver is anatomically divided into 4 major lobes, which are functionally divided into 8 segments based on vascular supply. Each segment is supplied by branches of the portal vein and hepatic artery (inflow), hepatic vein (outflow), and bile duct (Juza & Pauli, 2014). Within each segment, at a microscopic level the liver is organized into repeating units called **liver lobules** (Ben-Moshe & Itzkovitz, 2019).

2.1 Micro-structural organization of the liver lobule

The liver lobule is a hexagonal area of tissue, and contains a single central vein which drains blood away from the liver lobule (via the hepatic vein which eventually drains into the inferior vena cava). At each of the 6 corners of the hexagon lies a portal triad, consisting of three structures: a portal vein (carrying nutrient-rich blood from the intestines), an hepatic artery (carrying oxygenated blood from the coeliac trunk off the abdominal aorta), and a bile duct (which drains bile, an excretory product) (Ben-Moshe & Itzkovitz, 2019; Juza & Pauli, 2014; Ober & Lemaigre, 2018). This organization is demonstrated in Fig. 1A and B.

The inflow blood vessels (portal vein/hepatic artery) and outflow blood vessel (central vein) are connected by specialized capillaries called sinusoids, which are low pressure channels that are highly porous due to fenestrations present in the endothelial cells (**liver sinusoidal endothelial cells, LSEC**), and the discontinuous or minimal basal lamina outside the vessels (Poisson et al., 2017). Directly adjacent to the sinusoids is the perisinusoidal (interstitial) space, also called the space of Disse, containing **hepatic stellate cells**, which are liver-specific stromal cells that store lipids and vitamin A and act as pericytes during normal homeostasis, but can become activated and transition into myofibroblast-like cells during liver injury. The perisinusoidal space also contains **Kupffer cells**, liver-resident macrophages that are involved in immune surveillance and the clearance of pathogens, iron metabolism, and the innate immune response to liver injury (Wen, Lambrecht, Ju, & Tacke, 2021).

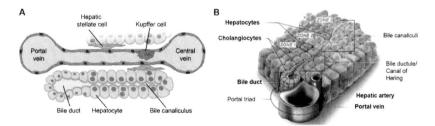

Fig. 1 Microanatomy of the liver lobule. (A) A section of the liver lobule is shown, extending from the central vein (in the center of the lobule) to the portal vein, located in one of the six hexagonal corners of a liver lobule (Ober & Lemaigre, 2018). (B) This section of the liver lobule shows liver zonation, where zone 1 lies near the portal triad (hepatic artery, portal vein, bile duct), zone 3 lies near the central vein (furthest away from the portal triad), and zone 2 is an intermediate zone (Raven & Forbes, 2018).

Adjacent to the perisinusoidal space are **hepatocytes**, the main paren-chymal cells (equivalent to 70% of all cells in the liver) (Blouin, Bolender, & Weibel, 1977; Si-Tayeb, Lemaigre, & Duncan, 2010), which are arranged in cords (or plates) radiating from the central vein to the perimeter of the lob-ule. Hepatocytes are polarized epithelial cells, expressing different receptors, transporters, and structures on their apical and basal surfaces (Gissen & Arias, 2015). Between two sinusoids lie two plates of hepatocytes, with the basal surface of each plate abutting the sinusoids on either side. The apical surface of each plate faces each other with a thin 1–2 μm space in between, forming a channel, as hepatocytes within each sheet form tight junctions between adjacent cells at their apical surface to form a continuous barrier. These channels are called bile canaliculi, because they transport bile produced and modified by hepatocytes as it flows from the perivenous region (near the central vein) to the periportal region (Banales et al., 2019; Boyer & Soroka, 2021). Toward the periportal area, the dual plates of hepatocytes transition into **cholangiocytes**, which form a bile duct that is continuous with the bile canaliculi. The transition area, called the Canal of Hering, is lined by periportal hepatocytes at its proximal end, which transition into a progenitor-like subset of cholangiocytes which have bipotential capacity to form hepatocytes and cholangiocytes, and distally form mature cholan-giocytes that are part of the intrahepatic bile duct system (Banales et al., 2019; Campana, Esser, Huch, & Forbes, 2021).

In terms of other cell types present in the liver, the large vessels (central vein, hepatic artery, portal vein) all contain **generic (non-LSEC) endo-thelial cells**, **pericytes** and **vascular smooth muscle cells**, and the portal triads are associated with **portal fibroblasts** which are part of the liver mes-enchyme (Dobie et al., 2019; MacParland et al., 2018; Ramachandran et al., 2019). A diverse population of immune cells (**natural killer cells, lympho-cytes, neutrophils**) are also present, although **Kupffer cells** represent the major immune cell type in the liver (Aizarani et al., 2019; MacParland et al., 2018; Ramachandran et al., 2019). Additionally, the liver contains a small number of **lymphatic vessels** (Frenkel et al., 2020), as well as **autonomic nerves** (branches of the sympathetic coeliac plexus and parasympathetic vagus nerve), which follow the course of the portal triads (Mizuno & Ueno, 2017).

2.2 Liver zonation

The liver lobule itself is not a homogeneous unit, but organized into three "zones," originally defined as three regions of hepatocytes with differing metabolic functions due to the gradient of oxygen, nutrients, metabolites, and signaling factors that arise as blood travels in one direction from the periportal area to the central vein, and as bile travels in a counter-current direction toward the periportal area (where the bile duct lies) (Raven & Forbes, 2018) (Fig. 1B). **Zone 1** is defined as the periportal area, where functions such as gluconeogenesis, cholesterol biosynthesis, fatty acid β-oxidation and ureagenesis predominate. **Zone 3** is the perivenous area (surrounding the central vein), where glycolysis, bile acid biosynthesis, glutamine synthesis, and xenobiotic metabolism occur. **Zone 2** is the transitional mid-area between zones 1 and 3, and localized functions include iron homeostasis, and modulation of insulin-like growth factors. Recent evidence suggests Zone 2 hepatocytes are the main contributors to homeostatic maintenance of the parenchyma during normal cell turnover (Ben-Moshe et al., 2019; Halpern et al., 2017; Wei et al., 2021).

Increasing evidence suggests that zonation is a phenomenon not just restricted to hepatocytes, but also applies to non-parenchymal cells such as hepatic stellate cells (Dobie et al., 2019; Payen et al., 2021), liver sinusoidal endothelial cells (Halpern et al., 2018; MacParland et al., 2018), and even the immune cells (Gola et al., 2021). Concurrently, the niche factors that regulate zonation have been investigated, revealing that non-parenchymal cells contribute directly to zonation, by releasing Wnt/β-catenin signals in the perivenous zone 3 area and Notch signals in the periportal zone 1 area (Brosch et al., 2018; Planas-Paz et al., 2016). Other important factors include gut-derived cues such as nutrients and commensal microbes, as well as pancreatic glucagon which are in high concentration in the peri-portal area where it enters the liver lobe (Cheng et al., 2018; Gola et al., 2021). These findings suggest that the entire liver lobule including its cellular and microenvironmental constituents are zonated, which collectively contributes to its functional zonation.

2.3 Extracellular matrix

The extracellular matrix (ECM) of the liver forms an important component of its microenvironment, although it only comprises a small percentage of the total liver volume. It regulates the function and structural organization of cells by facilitating cellular cross-talk with its neighbors and surrounding ECM, serves as a reservoir for signals such as growth factors and cytokines, and also plays a major role in the pathophysiological processes associated with liver injury and regeneration. The stromal compartment of the liver consists of the stellate cells in the space of Disse, periportal fibroblasts located in the vicinity of the portal triads, and pericytes associated with large blood vessels (Dobie et al., 2019; Strauss, Phillips, Ruggiero, Bartlett, & Dunbar, 2017). Although stromal cells are considered to be the main contributors of ECM, there is some evidence that other liver cells can also produce ECM, including hepatocytes (Ogata, Mochida, Tomiya, & Fujiwara, 1991; Tseng et al., 1982), cholangiocytes (Verhulst, Roskams, Sancho-Bru, & van Grunsven, 2019), LSECs (Bhandari et al., 2020) and liver progenitor cells (Berardis et al., 2014; Lorenzini et al., 2010).

Broadly, the ECMs in the liver can be categorized as **collagens**, **glyco-proteins**, and **proteoglycans**. The most abundant is **collagen I**, which is the main structural ECM in the liver, and other abundant components include **collagen IV** and **fibronectin**. In recent years, proteomic analysis has been used to define the human liver "matrisome," which is now recognized as a highly complex and dynamic tissue component consisting of over 150 ECM and ECM-associated proteins (Arteel & Naba, 2020; Daneshgar et al., 2020). The main deposits of ECM in the liver are the fibrous connective tissue capsule surrounding each of the liver lobes called the Glisson's capsule, the portal triads and central veins, and in the perisinusoidal space of Disse which serves as the interstitial compartment (Zhu, Coombe, Zheng, Yeoh, & Li, 2012). Each of these areas contain a slightly different composition of ECM, for example structural collagens (such as collagen I) are more abundant in the external connective tissue capsule and within the large blood vessels, whereas basement membrane ECM components are more dominant in the interstitial space and particularly in the periportal region where they contribute to the progenitor-cell niche (Lorenzini et al., 2010; Martinez-Hernandez & Amenta, 1993). Most epithelial tissues contain a well organized basement membrane supporting the epithelium, consisting of three structural layers (*lamina lucida*, *lamina densa*, and *lamina fibroreticularis*), which separates the epithelium from the stroma and vasculature (Mak & Mei, 2017). However, in the liver there are no definitive basement membrane structures, except in the major blood vessels (hepatic and portal arteries, central vein) and bile ducts. The basement membrane of these major structures consists of laminin, collagen IV, entactin, heparan sulfate proteoglycan, perlecan, and agrin (Martinez-Hernandez & Amenta, 1993; Roskams, Rosenbaum, De Vos, David, & Desmet, 1996; Tatrai et al., 2006). Additional ECMs found in the portal tract include non-basement membrane components such as collagen I, III, V, VI, fibronectin, and elastin (Couvelard et al., 1998; Martinez-Hernandez & Amenta, 1993). Within the liver parenchyma, loose interstitial matrix is found in the space of Disse, consisting of abundant fibronectin and collagen I, with small amounts of other collagens (types III, IV, V, and VI), heparan sulfate proteoglycan, tenascin, and little or no laminin (Couvelard et al., 1998; Martinez-Hernandez & Amenta, 1993; Zhu et al., 2012). Other elements of the ECM which have not been associated with a particular liver compartment include structural glycoproteins (undulin, nidogen, tenascin and osteonectin, and vitronectin), proteogylcans (biglycan, decorin, lumican, aggrecan, syndecan), glycosaminoglycans (heparan sulfate, chondroitin-4-sulfate, chondroitin-6-sulfate, dermatan sulfate), and hyaluronan (Arteel & Naba, 2020; Daneshgar et al., 2020; Gressner, Krull, & Bachem, 1994; Martinez-Hernandez & Amenta, 1993).

In line with the concept of zonation in the liver lobule, there is an ECM gradient in the liver. In zone 1, the periportal region consists of laminins and collagens III and IV, which are similar to the foetal liver ECM and maintains the liver progenitor cells located in this niche (Lorenzini et al., 2010; McClelland, Wauthier, Uronis, & Reid, 2008). In contrast, in the normal adult liver the pericentral zone 3 consists predominantly of collagen I and fibronectin, which modulates cell behavior by supporting differentiation and maintaining growth arrest (McClelland, Wauthier, Uronis & Reid, 2008).

3. Cell sources to recreate the liver niche

3.1 Primary cells

All the major cell types from the liver can be isolated, although some cell types are more restricted in their proliferative capacity and tend to dedifferentiate in culture. A major limitation in the use of primary human cells is the lack of healthy human tissues available for research. **Hepatocytes** are notoriously sensitive cells that rapidly decrease in viability during tissue transport and cell isolation, attach poorly to cell culture surfaces, and do not proliferate in culture (Bhogal et al., 2011; Kawahara et al., 2010). They rapidly dedifferentiate within 24h post-isolation, and do not last more than 7 days in monolayer culture (Heslop et al., 2017). They are also easily damaged during cryopreservation, resulting in significantly decreased viability, attachment and function (Terry, Dhawan, Mitry, Lehec, & Hughes, 2010). This makes hepatocytes an expensive, and often an unreliable source of parenchymal cells.

Cholangiocytes are more reliably isolated and readily expand in culture, and can also form cystic and ductular structures in 3D culture (Lewis et al., 2018; Sampaziotis et al., 2017; Tysoe et al., 2019). **Liver sinusoidal endothelial cells** are similar to hepatocytes in that they cannot be expanded in culture as they rapidly dedifferentiate (Xie et al., 2013), although co-culture with hepatocytes may prevent dedifferentiation (Bale et al., 2015). **Kupffer cells** can be isolated and there is some evidence of proliferation in culture, but the main issue with these cells is the relative impurity of cells, as endothelial cells are often a contaminant in isolations (Lynch et al., 2018; Zeng et al., 2013). **Hepatic stellate cells** have also been isolated and cultured, but the difficulty is isolating and maintaining quiescent cells, rather than activated cells which are myofibroblasts and much more proliferative (Zhai et al., 2019). In general, methods for hepatocyte and cholangiocyte isolation and culture are well-established, whereas the methods for other non-parenchymal cells lack standardization and face challenges in achieving adequate purity and maintaining cell quality.

3.2 Adult stem/progenitor cells

Liver progenitor cells can be isolated from adult liver tissue, expanded in culture, and be differentiated into both hepatocytes and cholangiocytes (Huch et al., 2015; Schneeberger et al., 2020). Liver progenitor cells can also be artificially induced in culture from primary hepatocytes, by adding small molecules that manipulate three pathways: the rho-associated protein kinase (ROCK), transforming growth factor β (TGFβ) and glycogen synthase kinase-3 (GSK3) pathways (Katsuda et al., 2017). Although progenitors derived from the liver are the most obvious choice to generate liver cells, trans-differentiation of extra-hepatic stem/progenitor cells also present alternative (but somewhat esoteric) sources. These include **umbilical cord-blood derived stem cells** (Tang, Zhang, Yang, Chen, & Zeng, 2006), **amnion epithelial cells** (Marongiu et al., 2011), **adipose tissue-derived mesenchymal stem cells** (Banas et al., 2007) and **pancreatic progenitor cells** (Gratte et al., 2018).

A major advantage of using a progenitor-based system is the ability to expand primary cells and terminally differentiate them on demand. However, current challenges include the development of differentiation protocols which produce a high yield of desired cells, and addressing safety concerns regarding the risk of tumor formation by liver progenitor cells due to their ability to act as cancer stem cells (Yamashita & Wang, 2013).

3.3 Pluripotent stem cells

Pluripotent stem cells have the capacity to differentiate into any cell arising from the three germ layers (endoderm, ectoderm, mesoderm). They can either be embryonic stem cells, which are derived from the inner cell mass of a blastocyst (Thomson et al., 1998), or be artificially generated in the laboratory by genetically reprogramming a somatic cell using the pluripotency factors *Oct4, Sox2, Klf4,* and *cMyc* (Takahashi et al., 2007).

Human embryonic stem cells (hESCs) have been used to generate hepatocytes (Ang et al., 2018), cholangiocytes (Dianat et al., 2014), and liver sinusoidal endothelial cell-like derivatives (Arai et al., 2011; Gage et al., 2020). However, hESCs are controversial due to the requirement for human embryos for their production, resulting in ethical and legal restrictions and a limit to the number of hESC lines available (Gopalan, Nor, & Mohamed, 2018).

Recently, **human induced pluripotent stem cells (hiPSCs)** have become much more widely used because they offer the same pluripotent potential as hESCs, but can be easily generated from a wide variety of patients using cell samples obtained via minimally invasive techniques, including from blood, urine or skin (Gaignerie et al., 2018; Kao et al., 2016; Takahashi et al., 2007). Most hiPSC-based studies have focused on the differentiation of hepatocytes and cholangiocytes, with many studies showing protocols for hESC can be applied to hiPSC (Ang et al., 2018; Baxter et al., 2015; Dianat et al., 2014; Sampaziotis et al., 2015). LSEC-like cells have been generated from hiPSC, but current evidence suggests these cells need to be transplanted *in vivo* into a mouse xenograft model to achieve maturity (De Smedt et al., 2021; Koui et al., 2017). Kupffer cells and hepatic stellate cells are the least studied cell types in terms of hiPSC differentiation so far, with only one study for each cell type (Coll et al., 2018; Tasnim et al., 2019).

Given the intensity of research in this area, it is expected that the fidelity of hiPSC-derived liver cells to their primary counterparts will increase over time as protocols for hiPSC differentiation are optimized and standardized. This will ultimately lead to a diverse, accessible, and cheaper supply of personalized cells for bioengineered liver platforms, that overcomes many of the challenges associated with primary cells.

3.4 Cell lines

Immortalized human cell lines exist for almost all liver cell types, except Kupffer cells. Common human hepatocyte cell lines include **HepG2, C3A, Huh7, Hep3B,** and **THLE-2** (Nwosu et al., 2018; Sefried, Haring, Weigert, & Eckstein, 2018; Sison-Young et al., 2015). The majority of these lines are derived from cancer specimens, are tumorigenic, and are phenotypically different from primary hepatocytes. **HepaRG** is a cell line established from hepatocellular carcinoma, but is non-tumorigenic and is considered a bipotent liver progenitor cell line because it can differentiate into both hepatocytes and cholangiocytes using specific protocols (Parent, Marion, Furio, Trépo, & Petit, 2004).

H69 is a well-characterized human cholangiocyte line derived from normal human cholangiocytes (Tabibian et al., 2014), and **HuCCT1** is a cholangiocyte line derived from cholangiocarcinoma (Miyagiwa, Ichida, Tokiwa, Sato, & Sasaki, 1989). There are several human liver sinusoidal endothelial cell (LSECs) lines, including **SK-Hep-1** (Sue, Hal, Jim, Taylor, & Darlington, 1992), **TRP3** (Parent et al., 2014), and **HLSEC/ciJ** (Zhu et al., 2018). However, these lines are morphologically and phenotypically different to highly differentiated primary LSECs, and are therefore not very widely used. **LX-1** and **LX-2** are the most commonly used human hepatic stellate cell lines, but express the phenotype of activated cells (Xu et al., 2005). This limits the application of LX-1 and LX-2 in modeling the normal homeostatic liver micro-environment, where quiescent hepatic stellate cells dominate.

Recently, a new method of immortalization has been described, which involves the transduction of proliferating-inducing factors (human papilloma virus genes E6/E7) into cells, to produce highly proliferating cells even if they are normally minimally proliferative in culture. Using this technology, cell lines from human hepatocytes and liver sinusoidal endothelial cells have been created (Ramachandran et al., 2015; Sison-Young et al., 2015). It is also possible to apply this to any other cell type, although whether this process reliably maintains the cell's original phenotype is still unclear.

Ultimately, cell lines are cheap, readily accessible and have minimal requirements for cell culture, so are useful for the initial establishment and characterization of liver models. However, their use in modeling complex liver biology is restricted because they only partially reflect the phenotype of their native tissue-derived counterparts.

4. Extracellular matrix and scaffolds

Extracellular matrix (ECM) derivatives and **fabricated scaffolds** are two important components routinely used to recreate the liver niche, often in combination. ECMs are used to provide a microenvironment that includes biological (ECM proteins, associated growth factors and attachment molecules) and mechano-physical (environmental stiffness) cues (Hussey, Dziki, & Badylak, 2018). This regulates the attachment, differentiation, and tissue assembly of cells. ECMs may be supplied as a cell culture surface coating, a thick hydrogel that allows cells to be embedded, or as a scaffold. A scaffold is a physical structure such as a mesh or sponge, which contains channels or pores that cells can be seeded into, and physically supports their assembly into 3D constructs (O'Brien, 2011). Scaffolds can be derived from ECMs (such as decellularized tissue), synthetic material, or a combination of both.

Biologically derived scaffolds and matrices can be sterilized and provide a 'native-like organ' structure with bio-active molecules, and are more likely to be biocompatible with cells. They are generally not immunogenic, have low toxicity for cells (Hosseini et al., 2019), but require donor tissue, and the conditions under which ECM-derived matrices are produced is not always standardized (Croce, Peloso, Zoro, Avanzini, & Cobianchi, 2019). Biological materials are also prone to batch variation, and if animal components are used this is often associated with ethical and safety concerns (Ye, Boeter, Penning, Spee, & Schneeberger, 2019).

Synthetic scaffolds and matrices have the advantage of being economical and relatively easy to produce with tunable properties such as stiffness, degradation, and hydrophobicity (Perez, Jung, & Kim, 2017). Their defined components and constitution means they can be easily standardized, facilitating regulatory approval for clinical use (Ye et al., 2019). They also have no risk of biological pathogens, can be easily sterilized, and do not require donor tissue. However, their disadvantages include the lack of natural cell binding sites or bioactive cues and intrinsic hydrophobicity, leading to reduced biocompatibility (Perez et al., 2017). Their relative simplicity also does not account for the liver's complex microenvironment (Croce et al., 2019). Additionally, synthetic materials may provoke a foreign body immune response (Chen, Wang, & Xiao, 2020), and leach toxic products during degradation (Cao et al., 2006; Perez et al., 2017).

It is desirable that scaffolds and matrices developed for 3D culture of any tissue or organ duplicate the stiffness, topography and biochemistry of the native organ ECM (Liaw, Ji, & Guvendiren, 2018), however these principles have only been replicated in part or not at all in many 3D bioengineered liver models. In addition to providing biocompatibility and supporting self-assembly, hepatic cell differentiation and cell polarity (Ye et al., 2019), the ideal scaffold/matrix: 1) facilitates cell attachment, 2) supports a nutrient-rich milieu which promotes cell viability, and 3) degrades over time at approximately the same rate that new ECM is laid down, and, cell proliferation occurs (Hosseini et al., 2019).

4.1 Cell culture surface coatings

The simplest use of ECMs to modulate the cell microenvironment is to use individual ECM components as a coating material for cell culture surfaces, for monolayer culture. This technique involves applying a thin layer of ECM on cell culture plastic surfaces, by overlaying a solution containing the ECM component and allowing a period of incubation, after which the solution may be discarded or allowed to air-dry. **Collagen I**, usually from rat tail, is the most common reagent used in liver culture, and has been shown to support the attachment of hepatocytes (Bhogal et al., 2011; Watanabe et al., 2016), cholangiocytes (Auth et al., 2001), liver sinusoidal endothelial cells (Xie et al., 2013), Kupffer cells (Tasnim et al., 2019), and hepatic stellate cells (Olsen et al., 2011). Other reagents that have been used are **laminin,** which supports both hepatocytes and cholangiocytes (Kanninen et al., 2016; Takayama et al., 2016; Watanabe et al., 2016), and **collagen III/IV** which supports liver progenitor cells (McClelland, Wauthier, Zhang, et al., 2008). **Fibronectin** is an ECM coating routinely used for the culture of a variety of endothelial cells, including liver sinusoidal endothelial cells (Dingle et al., 2018).

4.2 Decellularized ECM matrices

More complex ECM derivatives can be derived directly from liver tissue by removing its cellular content (Fig. 2A and B). Decellularized liver matrices have a number of advantages in liver tissue engineering. They create an acellular scaffold very similar in chemical composition and internal physical micro-geometry to the native liver. Thus decellularized liver ECMs are often considered more appropriate for liver tissue engineering than synthetic or biological scaffolds/matrices. This environment provides an appropriate extra-cellular matrix to guide cell attachment and engraftment, native micro-anatomical architecture, and biochemical cues for cell maturation and function.

Decellularized liver ECMs can be used to generate **cell culture coatings for 2D culture**, or can be used as **hydrogels** (Fig. 2D–G) or a **scaffold for 3D culture** (Hussein, Park, Yu, Kwak, & Woo, 2020; Mazza et al., 2017; Park et al., 2016; Wang et al., 2011). Alternatively, the entire liver or specific lobes of liver tissue can be regenerated by **repopulating**

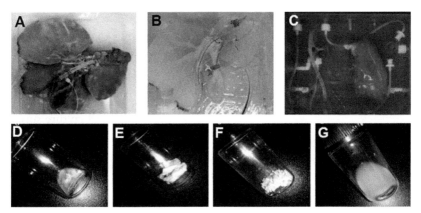

Fig. 2 Decellularized liver ECMs. (A) A human liver prior to decellularization, (B) the human liver at the end of decellularization, where it becomes translucent due to the loss of blood and cells, (C) a decellularized human liver lobe connected to a perfusion system and infused with cells via its blood vessels to repopulate the ECM scaffold. (D) Decellularized liver ECM, (E) lyophilized decellularized liver ECM, (F) liver ECM milled into a powder form, and (G) powdered liver ECM reconstituted into a hydrogel. *Panels (A–C): From Willemse, J., Verstegen, M. M. A., Vermeulen, A., Schurink, I. J., Roest, H. P., van der Laan, L. J. W., et al. (2020). Fast, robust and effective decellularization of whole human livers using mild detergents and pressure controlled perfusion.* Materials Science and Engineering: C, 108, 110200. *doi:10.1016/j.msec.2019.110200. Panels (D–G): From Hussein, K. H., Park, K. M., Yu, L., Kwak, H. H., & Woo, H. M. (2020). Decellularized hepatic extracellular matrix hydrogel attenuates hepatic stellate cell activation and liver fibrosis.* Materials Science & Engineering. C, Materials for Biological Applications, 116, 111160. *doi:10.1016/j.msec.2020.111160.*

decellularized liver matrices with viable cells (Fig. 2C) (Baptista et al., 2011; Mazza et al., 2015; Shaheen et al., 2020; Takeishi et al., 2020). This is a new and expanding field with new reports of technique improvements and revised applications published regularly. More recent applications of decellularized liver ECM include the manufacture of bio-inks for use in bio-printing, either to fabricate ECM-based scaffolds (Grant, Hallett, Forbes, Hay, & Callanan, 2019), or to bio-print 3D tissues using cell-laden bio-inks (Lee et al., 2017; Yu et al., 2019).

Decellularization techniques generally involve removal of the living cells of an organ by either physical, chemical or enzymatic agents that induce cell necrosis and lysis (McCrary, Bousalis, Mobini, Song, & Schmidt, 2020), usually by intravascular infusion of detergents and/or enzymes and subsequent washing. The resulting material retains the non-living, three-dimensional connective tissue structural framework of the organ, which includes natural cell binding sites which assist in cell attachment, proliferation and differentiation (Badylak, Taylor, & Uygun, 2011). These structures retain various basic ECM components including collagens, glycoproteins (such as fibronectin and laminin) and glycosaminoglycans (GAGs), and a cohort of signaling and growth factors specific to that organ (Badylak et al., 2011). A significant advantage of using decellularized organ/tissue matrices is their reduced immunogenicity due to removal of the cellular fraction which can provoke an immune response if present (Fishman et al., 2013). It should also be noted that damage to the native ECM may also occur in the decellularization process, and this is dependent on the technique used (Faulk et al., 2014; Ren et al., 2013).

Whole liver decellularization was first reported in 2010 using rodent livers (Shupe, Williams, Brown, Willenberg, & Petersen, 2010; Uygun et al., 2010), but was quickly replicated in larger animals including rabbits and pigs (Baptista et al., 2011; Coronado et al., 2017; Mirmalek-Sani, Sullivan, Zimmerman, Shupe, & Petersen, 2013; Struecker et al., 2015; Takeishi et al., 2020; Yagi et al., 2013). These studies confirmed that dec-ellularized liver matrices were acellular with complete removal of DNA content, and maintained the liver's ECM components and biliary/vascular framework. There was also some evidence that growth factors such as HGF, bFGF, EGF, VEGF and IGF-1 were still present (Coronado et al., 2017; Yagi et al., 2013).

The first attempt to decellularize human livers was reported in 2015 (Mazza et al., 2015), using whole or partial (split) human livers that were not suitable for transplantation. The livers were perfused with saline to remove blood, freeze-thawed to promote cell rupture, and decellularized with the retrograde intravascular perfusion of sodium dodecyl sulfate (SDS), peracetic acid, Triton X-100, and ethanol. This protocol took up to 6 weeks for whole livers, and 14 days for partial livers. This method was revised in 2017 to a more rapid protocol, involving the cutting of liver into cubes and the use of high oscillation shear stress with detergent to pro-mote decellularization, which reduced processing time to about 24 h (Mazza et al., 2017). Since then, other groups have reported decellularized human liver matrices (Grant et al., 2019; Willemse et al., 2020). A major limitation in the use of human livers is the logistical difficulty in obtaining and handling the large volumes of human liver tissue required, considerable heterogeneity between patients, and that most livers donated for research are either dis-eased or non-viable due to prolonged ischemia, which may negatively impact the ECM composition (Mattei, Magliaro, Pirone, & Ahluwalia, 2017; Mazza et al., 2015, 2017). In this respect, pig livers have been of major interest, because of the relative similarity in organ size and structure to humans (Nykonenko, Vávra, & Zonča, 2017), the ability to harvest healthy livers, and the interspecies biocompatibility of ECMs (McClelland, Wauthier, Zhang, et al., 2008; Takeishi et al., 2020).

4.3 Hydrogels

Hydrogels are either solubilized biological ECMs or synthetic cross-linked polymer networks, which can absorb and retain large amounts of water (Caliari & Burdick, 2016). They can polymerize and solidify to form a soft 3D construct under defined conditions such as temperature, pH, or the presence of a specific enzyme. Their soft consistency is closer to liver tissues than many other scaffolds and matrices (Liaw et al., 2018), and they are easy to handle during application. However, they may not provide adequate mechanical strength and are prone to rapid degradation, and therefore chemical or physical cross-linking may be used to enhance hydrogels (Perez et al., 2017).

The use of hydrogels can be classified broadly into 4 applications, which are: (i) **2D monolayer cell culture,** where hydrogels are diluted and used to produce a thin surface coating (Bhogal et al., 2011), (ii) **3D sandwich culture**, where cells are overlaid onto a solidified hydrogel layer, then another hydrogel layer is formed on top of the cells (Bachour-El Azzi et al., 2015; Sellaro et al., 2010), (iii) **3D embedded culture**, where cells are suspended within a thick hydrogel construct to form aggregates such as spheroids or organoids (Huch et al., 2015; Sampaziotis et al., 2015), and (iv) **combination culture**, where 2D, sandwich or 3D-embedded culture is combined with other techniques such as microfluidic systems, perfusion bioreactors, and transwell culture inserts (Kang, Rawat, Cirillo, Bouchard, & Noh, 2013; Ramachandran et al., 2015).

Biological hydrogels include **collagen I** (Bachour-El Azzi et al., 2015)—a highly abundant structural ECM found in every organ across all animal species, **gelatin** (Chen et al., 2020)—a hydrolyzed form of collagen, **fibrin** (Stevens et al., 2017)—a liver-derived product of coagulation, **hyaluronan** (Chitrangi, Nair, & Khanna, 2017)—a widely distributed ECM polysaccharide, and **alginate** (Lau, Ho, & Wang, 2013)—a natural polysaccharide derived from brown algae.

Complex biological hydrogels comprising multiple ECM components include **liver ECM-derived hydrogels** (Hussein et al., 2020; Lewis et al., 2018), and **Matrigel™** (Ramachandran et al., 2015; Takebe et al., 2013). Matrigel is one of the most commonly used ECM derivatives, and is a hydrogel derived from mouse Engelbreth-Holm-Swarm sarcoma, a connective tissue tumor that produces abundant ECM. Matrigel is a very complex mixture with approximately 1800 proteins (Hughes, Postovit, & Lajoie, 2010), but its major components are laminin (60%), collagen IV (30%), entactin (8%), and perlecan (2–3%), which are all basement membrane ECMs (Aisenbrey & Murphy, 2020). The majority of seminal studies in liver bio-engineering have all used Matrigel, because it is highly efficient in promoting cell attachment, differentiation, and self-assembly into tissue structures (Hu et al., 2018; Huch et al., 2013, 2015; Sampaziotis et al., 2015; Takebe et al., 2013).

Biological matrices can be used in combination, such as **hyaluronan-collagen hydrogels** (Nishii et al., 2018), **hyaluronan-liver ECM hydrogels** (Deegan, Zimmerman, Skardal, Atala, & Shupe, 2015), and **hyaluronan-gelatin hydrogels** (Malinen et al., 2014). They can also be combined with synthetic materials, such as the combination of **laminin with synthetic polymer polyisocyanopeptides** (Ye et al., 2020), **hyaluronan-polyethylene glycol (PEG) hydrogels** (Christoffersson et al., 2018), and **gelatin-methacroloyl (GelMA)** (Cuvellier et al., 2021; Wu, Wenger, Golzar, & Tang, 2020). GelMA is commonly used in bio-printing, and is a very versatile, biocompatible and photo-crosslinkable biomaterial.

Synthetic hydrogels include **polyethylene glycol (PEG)**, which is widely used in liver studies (Li et al., 2014; Schepers, Li, Chhabra, Seney, & Bhatia, 2016; Underhill, Chen, Albrecht, & Bhatia, 2007), **poly(lactide-co-glycolide) acid (PLGA)** (Lee, Cho, Xiong, Glenn, & Frank, 2010), the peptide hydrogel **PuraMatrix™** (Wang, Nagrath, Chen, Berthiaume, & Yarmush, 2008) and **polyvinyl alcohol (PVA)** (Xia, Zhao, Feng, & Yang, 2020). Although many biomaterials-focused studies have been conducted using synthetic hydrogels, their lack of bioactivity and natural viscoelasticity are known negatives (Ye et al., 2019), and limits their use in advanced liver models that aim to faithfully recapitulate the biological complexity of the liver microenvironment.

4.4 Fabricated scaffolds

Scaffolds are more physically defined structures than hydrogels, and are designed and fabricated to have specific properties that provide mechanical support, allow cell attachment, and facilitate tissue assembly within gaps, pores, or channels (O'Brien, 2011). The following scaffold types have been used to create a liver tissue niche including: **microspheres** (that contain hollow cores where cells become embedded) (Siltanen et al., 2017), **fibrous meshes** (Bishi et al., 2016), **sponges** (Kukla, Stoppel, Kaplan, & Khetani, 2020), **lattices** (Ng et al., 2018), or **electrospun constructs** (Grant et al., 2019) which may contain **nanofibers** (Rajendran et al., 2017).

Common biological scaffolds used for liver tissue assembly include materials often used in hydrogels, such as **collagen** (Ruoß et al., 2018), **hyaluronan** (Teratani et al., 2005), **gelatin** (Hou & Hsu, 2020), and **alginate** (Dvir-Ginzberg, Gamlieli-Bonshtein, Agbaria, & Cohen, 2003). However, they are often either modified, or blended with other materials, and usually used at much higher concentrations than hydrogels, to fabricate a mechanically robust scaffold. Additional biological materials used for scaffolds to produce liver tissue include **silk** (Kukla et al., 2020)—derived from the secretions of silk worms and spiders, **cellulose** (Nugraha et al., 2011)—a plant-derived polysaccharide, and **chitosan** (Feng et al., 2009)—derived from shell fish.

Synthetic scaffolds used for assembly of liver tissue include those manufactured using **polyethylene glycol (PEG)** (Ng et al., 2018), **poly-L-lactic acid (PLLA)** (Török et al., 2011), **poly(l-lactide–co–glycolide) PLGA** (Brown et al., 2018), **polycaprolactone (PCL)** (Grant, Hay, & Callanan, 2017) and **polyurethane** (Linti et al., 2002). Scaffolds can also combine biological and synthetic materials, such as a **silk–liver ECM** (Janani & Mandal, 2021), **PLLA–liver ECM** (Grant et al., 2019), and **PLGA–collagen** (Brown et al., 2018) blend scaffolds. These biomaterials combine the advantages of both biological and synthetic components to increase biocompatibility, cellular function/assembly, and optimize degradation properties, while maintaining tunable characteristics and mechanical strength. Additionally, scaffolds can also be used in combination with hydrogels, which can be used to coat scaffold surfaces to enhance their biocompatibility (Ng et al., 2018), or as a carrier for cells that are seeded into the scaffold (Ijima, Hou, & Takei, 2010).

5. Bio-engineered platforms to recapitulate the liver niche

5.1 Two-dimensional (2D) platforms

A simplified liver niche can be created *in vitro* through the **co-culture of hepatocytes with non-parenchymal cells** as a monolayer. The aim of these early studies was to prolong the *in vitro* maintenance of primary hepatocytes, by using non-parenchymal cells as support cells. An ECM coating on the cultured surface is used to facilitate the attachment of cells, which is typically collagen I (DeLeve, Wang, Hu, McCuskey, & McCuskey, 2004; Morin & Normand, 1986; Uyama et al., 2002). Non-parenchymal cells used in these approaches include liver sinusoidal endothelial cells (Morin & Normand, 1986), hepatic stellate cells (Uyama et al., 2002), and extra-hepatic cells such as mesenchymal stem cells (Fitzpatrick et al., 2015) and fibroblasts (Michalopoulos, Russell, & Biles, 1979). These may be primary cells, or cell lines. The non-parenchymal cells provide paracrine support and heterotypic cell interactions, leading to stabilization of the mature hepatocyte phenotype in long-term culture, typically up to several weeks. Non-parenchymal liver cells have also been shown to promote the differentiation of embryonic stem cells into hepatocytes (Soto-Gutiérrez et al., 2007), and the co-culture of hiPSC-derived liver progenitor cells together with hiPSC-derived liver sinusoidal endothelial cells enhances the differentiation of both cell types (Danoy et al., 2021).

With increasing interest in non-parenchymal cells and their role in liver biology, studies have started to focus on how the highly specialized tissue-specific phenotype of these cells can be preserved in culture, and the interactions between specific cell types that contribute to organ function. Examples of such studies include co-culture systems where liver sinusoidal endothelial cells can be maintained when co-cultured with hepatocytes, but dedifferentiate when cultured alone (DeLeve et al., 2004). Hepatocyte-Kupffer cell cultures have been used to probe the induction of immune response that occurs in drug-induced hepatotoxicity and infection (Rose, Holman, Green, Andersen, & LeCluyse, 2016), as well as their combined role in iron metabolism (Sibille, Kondo, & Aisen, 1988). Interactions between non-parenchymal cells have also been studied, for example the regulation of hepatic stellate cell activation by liver sinusoidal endothelial cells (DeLeve, Wang, & Guo, 2008).

Early techniques in co-culture studies involved the mixing of different cell types in specific ratios as a cell suspension, then plating the mixed culture as a monolayer (Morin & Normand, 1986), or adding hepatocytes onto an already growing monolayer of non-parenchymal cells (Uyama et al., 2002). A rudimentary self-assembly process produced islands of hepatocytes surrounded by support cells (Morin & Normand, 1986; Uyama et al., 2002). However, this process is unreliable, leading to varying degrees of island formation even within the same culture, and this does not occur if cells are plated at low density (Morin & Normand, 1986; Rose et al., 2016). Within islands, hepatocytes in the periphery benefit from heterotypic interactions with non-parenchymal cells, whereas cells in the center of particularly large islands are bereft of any direct heterotypic interactions (except paracrine factors in the media).

To exert more control on the spatial distribution of different cell types, **micropatterned co-culture** has been used. This creates hepatocyte islands of specific size and distribution by micropatterning islands of collagen-coating onto a cell culture surface using lithographic techniques, then plating hepatocytes to allow selective attachment to the collagen (March et al., 2015). Once hepatocyte islands are established, non-parenchymal cells such as endothelial cells and fibroblasts are seeded to fill the gaps between hepatocyte islands (Ware, Durham, Monckton, & Khetani, 2018). Such micropatterned co-culture systems have also been used to enhance the differentiation of hiPSC-derived hepatocytes through co-culture with fibroblasts (Ware, Berger, & Khetani, 2015).

Another 2D culture method is the use of **transwell cultures**. These are porous membranes that can be placed into multi-well cell culture plates, and are suspended above the bottom of the plate using a supportive device that hangs over the top edge of the wells. This allows different cell types to be indirectly "co-cultured" as separate mono-layers in the same well, either on top (on the transwell membrane) or bottom (on cell culture plastic). Because the membrane pores allow media and molecular exchange but do not allow cell migration between the separate compartments, co-culture systems allow the study of paracrine influences between cell types. Hepatocytes have been co-cultured in separate layers with mesenchymal stem cells (MSCs) to establish the paracrine effect of MSCs on hepatocytes (Fitzpatrick et al., 2015), and the influence of direct and indirect cell contacts on the maintenance of liver sinusoidal endothelial cell co-cultures have also been studied (DeLeve et al., 2004).

Additionally, monolayer cells cultured on a transwell membrane can be exposed to different types of media on their apical and basal surfaces. This results in different transporters and structures (such as microvilli) to be expressed on the apical and basal surface of hepatocytes, a phenomenon called polarization, and this allows the unidirectional transport of substances such as nutrients, drugs, or bile acids. This is lost in monolayer culture on plastic surfaces, but hepatocytes become polarized when cultured on transwell membranes (Dao Thi et al., 2020). A nutrient rich media can be supplied underneath the transwell (mimicking nutrient rich blood flowing underneath the basal surface of hepatocytes *in vivo*), while a minimally-enriched media can be supplied over the cells, mimicking the bile canaliculi which are excretory channels. These transwell methods consider the spatial organization of hepatocytes in the liver, where they are typically not heterogeneously mixed or in direct contact with non-parenchymal cells, but separated into compartments.

5.2 Three-dimensional (3D) platforms

Monolayer culture on a flat plastic surface poorly replicates the *in vivo* microenvironment of cells, which exist in a 3D spatial organization where dynamic cell-cell and cell-ECM interactions create tissue structures with a complex microenvironment. These limitations are addressed in multicellular constructs in a 3D configuration. The prototype models for 3D liver cell culture were **3D collagen embedding** and **collagen sandwich** methods for the culture of primary hepatocytes to prolong their *in vitro* lifespan beyond several days (Dunn, Yarmush, Koebe, & Tompkins, 1989; Gomez-Lechon et al., 1998). Similar embedding and sandwich methods were used with **Matrigel**, which was found to be more effective than collagen (Moghe et al., 1996), likely due to its complex mixture of ECMs. Hepatocytes cultured in these 3D hydrogels not only maintained viability and function, but structurally organized to become polarized, form bile canaliculi, and cell aggregates (Coger, Toner, Moghe, Ezzell, & Yarmush, 1997; LeCluyse, Audus, & Hochman, 1994; Liu et al., 1999; Moghe et al., 1996). Matrigel was also found to enhance the biliary differentiation of liver progenitor cells, which formed biliary cysts and ductular structures (Tanimizu, Miyajima, & Mostov, 2007). Interconnected bile duct structures have also been formed using cholangiocytes embedded within liver ECM hydrogels (Lewis et al., 2018), and the use of complex, liver-specific ECM hydrogels as an alternative to Matrigel is likely to expand in the future. Additionally, a wide variety of hydrogels derived from biological and synthetic materials have been used to culture liver cells, which have been described in the previous subsection "**Hydrogels**." In addition or as an alternative to hydrogels, many 3D liver constructs involve the use of scaffolds, to construct a more rigid, mechanically robust structure than hydrogel-based constructs. The wide variety of scaffolds that are used for such endeavors have already been described in the previous subsection "**Fabricated scaffolds**."

There are also other methods that do not require the use of hydrogels, scaffolds, or matrices. This includes the **hanging drop technique**, where cells are suspended in a droplet placed on the lid of a petri-dish which is then overturned, and cells within the droplet aggregate in a gravity-dependent fashion (Hurrell, Ellero, Masso, & Cromarty, 2018). Gravity can also be used with cells seeded into **ultra-low attachment U-bottom 96-well plates**, where cells sediment and aggregate at the bottom of the wells (Bell et al., 2016), which can be aided by the addition of a thickening agent such as methylcellulose to the cell culture medium (Dingle et al., 2018; Yap et al., 2013, 2020). Cells can also be plated into a mold fabricated with **inverted pyramid microwells**, and centrifuged to pellet the cells and allow aggregation within the microwells (Fey & Wrzesinski, 2012). Spontaneous cell aggregation can also be achieved through agitation and swirling of the cell suspensions using **orbital shakers** (Weeks et al., 2013), **rocking platforms** (Brophy et al., 2009), **spinner flasks** (Schneeberger et al., 2020), and **rotating wall perfusion bioreactors** (Tostoes et al., 2012).

A common finding in these studies is that 3D culture enhances the viability and function of hepatocytes, regardless of the technique used. Cells aggregate within these structures and secrete their own ECM (Yap et al., 2013). Most studies use the term **spheroids** to describe the cell aggregates formed, because they are spherical in structure. These methods have also been used to incorporate non-parenchymal cells to form co-cultured spheroids, to further enhance the maturity of hepatocytes and add additional functions to the construct, such as the ability to respond to inflammatory stimuli (Bell et al., 2016; Li, Cao, Parikh, & Zuo, 2020) and trigger fibrosis (Coll et al., 2018; Mannaerts et al., 2020). Many different combinations of **co-cultured spheroids** have been reported, including hepatocytes from various sources (primary, cell line, hiPSC-derived) with **endothelial cells** (Inamori, Mizumoto, & Kajiwara, 2009; Lasli et al., 2019), **hepatic stellate cells** (Coll et al., 2018; Mannaerts et al., 2020), **Kupffer cells** (Bell et al., 2016; Li et al., 2020), **cholangiocytes** (Hafiz et al., 2021), **fibroblasts** (Lu et al., 2005), and **mesenchymal stem cells** (No et al., 2012).

Other methods of 3D culture include the stacking of different cell layers, often with a membrane used to separate layers, either within a transwell or a biochip within a microfluidic device (Kang et al., 2013; Rennert et al., 2015). The layers usually contain different liver cell types, and such methods may be called **layered co-culture**. This mimics the plate-like configuration of hepatocytes extending from the center of the liver lobule to each of its 6 hexagonal corners. Another method is the use of **3D bio-printing**, where liver cells are loaded into bio-inks created from biological or synthetic hydrogels that are laid down as droplets or extruded like toothpaste, which can be repeated in layers and other spatial configurations to generate a 3D construct (Ma et al., 2020). Bio-printing offers the opportunity to lay down cells in a specific configuration, and this allows the size and organization of the construct to be tailored. The immediate micro-environment of cells can also be manipulated using bio-inks with specific properties, such as native tissue ECM components, appropriate viscoelasticity and stiffness, and they can also be manipulated over time using methods such as photo-curing to polymerize bio-inks at a specific time (Gungor-Ozkerim, Inci, Zhang, Khademhosseini, & Dokmeci, 2018). Specific advantages of bio-printing include the ability to scale up production and reproducibility due to auto-mation. However, successful bio-printing also depends on optimizing the bio-ink for biocompatibility, and minimizing shear stress and production times which can be detrimental to cells (Blaeser et al., 2015). Bio-inks also need to be degraded over time so cells can remodel and interact to form a dynamic tissue structure. Bio-printing methods are also currently in low res-olution without the scope for fine details, although whether high resolution bio-printing is possible without compromising cell integrity as they pass through the printer nozzle is unclear. This is a rapidly evolving field and is poised to become useful in drug testing, where high throughput models are particularly important for drug screening and large studies.

5.3 Organoids

Organoids are formed by a 3D culture technique, the use of which has expo-nentially increased within the last decade. It refers to an organ-like structure generated through the aggregation of cells, usually in the presence of exog-enous ECM and a very complex nutrient-rich medium containing a cocktail of growth factors and small molecules (Kim, Koo, & Knoblich, 2020; Marsee et al., 2021). A major difference between spheroids and organoids is the level of self-assembly and complexity. Spheroids are generally made in simplified cell culture conditions without additional ECMs, using basic medium (Yap et al., 2013). These conditions are often suitable for cell lines, although primary cells can also be used. Spheroids are also relatively homog-enous in structure, with little evidence of sophisticated self-assembly even in co-cultured spheroids with more than one cell type (Bell et al., 2016; Coll et al., 2018).

In contrast, a key attribute of organoids is the remarkable level of self-assembly, which may be attributed to the complex protocols used which provide many morphogens that direct tissue assembly. Even if derived from a single cell type, cells within an organoid are heterogeneous, with structural organization such as the polarization of hepatocytes and the formation of bile canaliculi (Hu et al., 2018), or bile ducts (Tanimizu et al., 2021) and blood vessels (Takebe et al., 2013). Hierarchical organization of cells with epithelial plasticity or retention of a progenitor compartment means organoids can often be propagated and expanded in culture (Hu et al., 2018; Huch et al., 2015). This level of complexity results in organoids possessing many structural and functional features found in native liver, as well as transcriptomic similarities. The exact line between spheroids and organoids is still a matter of debate, and some groups still use these terms interchangeability (Marsee et al., 2021; Velasco, Shariati, & Esfandyarpour, 2020). However, as the complexity and biological relevance of organoids increases, the term spheroid is becoming less widely used.

There is a vast array of liver organoids, with hepatocytes and/or cholangiocytes being the major component (Marsee et al., 2021). In terms of primary cell-derived liver organoids, the first major studies were the isolation of liver progenitor cells from adult mouse and human livers, which self-assembled into complex organoids which could be expanded in culture, and differentiated into mature hepatocytes or cholangiocytes by manipulating the cell culture media (Huch et al., 2013, 2015). A major component of these protocols was modulation of the Wnt signaling pathway, which was shown to be a key mediator of progenitor cell expansion and hepatobiliary differentiation. Since then, many primary liver organoids have been developed, including **hepatocyte organoids** (Hu et al., 2018; Peng et al., 2018), **cholangiocyte organoids** (Rimland et al., 2021; Sampaziotis et al., 2017, 2021), and **co-culture organoids combining hepatocytes and cholangiocytes** (Tanimizu et al., 2021), and **hepatocyte, endothelial cell and mesenchymal stem cell organoids** (Enomoto et al., 2014). **Cell line-derived organoids** have also been shown to develop complex self-assembling structures when similar additives used in primary organoid cultures are applied (Ramachandran et al., 2015; Yap et al., 2020). Additionally, **liver cancer organoids** have been derived using similar techniques, from human hepatocellular carcinoma and cholangiocarcinoma samples (Broutier et al., 2017; Saito et al., 2019). This is a useful method to expand small tumor samples to form patient bio-banks, and is clinically relevant as tumor organoids show morphological and transcriptomic similarities to the original tumor specimen.

hiPSC-derived liver organoids are also widely reported. The first major study in this area was reported in 2013, using hiPSC-derived hepatocytes co-cultured with primary endothelial cells and mesenchymal stem cells

(Takebe et al., 2013). Since then, many different hiPSC-derived organoids have been developed, often using ingredients and protocols that have been used with primary liver organoids. This includes cholangiocyte organoids (Sampaziotis et al., 2015), hepatobiliary organoids (Wu et al., 2019), and increasingly complex organoids with different cell types all completely derived from hiPSC, such as endothelial cells, stromal cells, Kupffer cells, and hepatic stellate cells (Koui et al., 2017; Ouchi et al., 2019).

Some different types of adult and hiPSC-derived organoids and liver constructs are shown in Fig. 3.

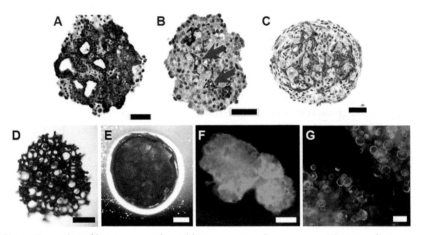

Fig. 3 Examples of liver organoids and bio-engineered constructs. (A) Mouse liver progenitor cells co-cultured with liver sinusoidal endothelial cells in an organoid format results in the robust self-assembly of hepatobiliary tissue, including pan-cytokeratin + (brown) ductular structures. (B) Mouse liver sinusoidal endothelial cells co-cultured in mouse vascularized hepatobiliary organoids assemble into a LYVE-1 + (brown) capillary network. (C) Human liver sinusoidal endothelial cells assemble into robust CD31 + (brown) capillary networks when cultured as a 3D organoid. (D) A 3 mm diameter circular polyurethane scaffold (NovoSorb® from PolyNovo Ltd., Melbourne, Australia) with interconnected pores. (E) The polyurethane scaffold can be seeded with human liver cells combined with a human liver ECM hydrogel to generate a human liver construct. (F) eGFP-expressing hiPSC-derived hepatocytes can be aggregated into self-assembling organoids by using an orbital shaker and growth-factor rich media. (G) eGFP-hiPSC-hepatocytes can alternatively be seeded into Matrigel and will form many small self-assembling organoids within Matrigel droplets. Figs. (C–G) supplied by the authors. Scale bars: 50 μm for (A–C), 500 μm for (D–F), and 200 μm for (G). *Panels (A and B) from Yap, K. K., Gerrand, Y-W., Dingle, A. M., Yeoh, G. C., Morrison, W. A., & Mitchell, G. M. (2020). Liver sinusoidal endothelial cells promote the differentiation and survival of mouse vascularised hepatobiliary organoids. Biomaterials, 251, 120091. doi:10.1016/j.biomaterials.2020.120091.*

5.4 Repopulated decellularized liver ECM scaffolds

Theoretically, constructs closest in structure to the native liver can be bio-engineered by repopulating a decellularized liver with the appropriate cells, so they use the native ECM as a guide to localize in the appropriate locations. This approach also offers the possibility of whole organ bio-engineering. However, it is still a relatively new field, and many methods need to be optimized, such as decellularization and repopulation techniques, the choice of cells, and bioreactor systems.

Methods can be grouped into either the repopulation of whole livers or liver lobes via direct injection of cells into the parenchymal ECM, the infusion of cells through the main blood vessels and bile duct, or the reseeding of decellularized liver ECM discs and cubes. Direct parenchymal injection is logistically easier, but leads to patchy repopulation and poor cell retention (Soto-Gutierrez et al., 2011). Cell infusion through the portal vessels (Fig. 2C) allows cells to travel throughout the entire tissue within the vasculature, and migrate into the parenchyma (Soto-Gutierrez et al., 2011; Takeishi et al., 2020; Willemse et al., 2020). This leads to much greater cell retention of up to about 90% of cells infused, particularly if cells are infused in a multi-step protocol to provide intervals between infusions to allow hepatocyte migration (Soto-Gutierrez et al., 2011). Multi-step protocols can also involve the initial infusion of hepatocytes to repopulate the parenchyma, followed by endothelial cells to line the vasculature (Uygun et al., 2010). A further step has also been reported, by infusing cholangiocytes into the bile ducts of a decellularized liver already repopulated with hepatocytes and endothelial cells (Takeishi et al., 2020).

Alternatively, decellularized liver matrices can be processed as small cubes (5 × 5 × 5 mm) or discs (5 mm diameter) to form scaffolds, and a small volume of a single cell suspension pipetted or injected into the scaffold, and cells allowed to attached over time in static culture (Mazza et al., 2015, 2017; Willemse et al., 2020). Cells may also be seeded by placing scaffolds into a pump-operated perfusion system, with cell suspensions flowing through a chamber containing the scaffold and attaching over time (Mazza et al., 2017). The size of constructs bio-engineered in this manner are generally limited by size of ECM cubes/discs used (usually small). There is also an assumption that seeded cells will migrate through the ECM network to localize and self-assemble in anatomically appropriate locations, for example hepatocytes within the parenchyma, cholangiocytes within bile ducts, and endothelial cells within blood vessels and sinusoids. Histological examination of the repopulated scaffolds generated in this manner indicates that such self-organization does not occur enough to convincingly recreate the native liver microarchitecture (Mazza et al., 2017, 2015; Willemse et al., 2020). However, there is some evidence that seeded human umbilical vein endothelial cells can migrate and localize in the luminal surface of major blood vessels, although they did not form a sinusoidal capillary network (Mazza et al., 2017; Verstegen et al., 2017).

Early studies have used decellularized rodent livers repopulated with rodent cells (Soto-Gutierrez et al., 2011; Uygun et al., 2010), but increasingly human cells are being used, either to repopulate a rodent decellularized liver (Takeishi et al., 2020) or even human liver (Mazza et al., 2017; Willemse et al., 2020). These studies have used cell lines (Mazza et al., 2017; Willemse et al., 2020), primary cells (Soto-Gutierrez et al., 2011; Uygun et al., 2010), and more recently, hiPSC-derived cells (Takeishi et al., 2020). However, repopulation requires fifty million to hundreds of millions of cells for even the size of a rodent liver (Soto-Gutierrez et al., 2011; Takeishi et al., 2020; Uygun et al., 2010), and 1 billion cells have been used for a pig liver (Yagi et al., 2013). This has limited the scale of experiments and certainly made repopulating a whole human liver an immense challenge. Nevertheless, the main use of a whole repopulated liver is for transplantation purposes, and for studies on liver biology a semi-populated liver lobe, rodent liver ECMs, and ECM discs/cubes may be sufficient.

5.5 Perfusion bioreactors and microfluidic systems

Perfusion bioreactors and microfluidic systems share the underlying concept that cell culture media can be circulated through a bioreactor chamber or a miniature chip device, to enhance nutrient and waste exchange, leading to enhanced cell proliferation, differentiation and viability in culture (Jang et al., 2019; Ramachandran et al., 2015; Rennert et al., 2015). This is an important microenvironmental consideration, because it mimics the dynamic circulatory exchange that occurs within tissues between its cells and blood in the vasculature, and it directly influences parameters such as nutrient and waste concentrations, oxygen tension, shear stress, temperature and pH (Underhill & Khetani, 2018; Vunjak-Novakovic, Ronaldson-Bouchard, & Radisic, 2021).

Perfusion bioreactors include rotating vessels with media being pumped through to maintain large volumes of suspended cell aggregates and constructs (Lu et al., 2016), or on a smaller scale individual chambers that enclose liver constructs that are perfused by an inflow and outflow tubing system which can be interconnected to different chambers (Ramachandran et al., 2015). Although scaled up production and long-term culture is made feasible by perfusion bioreactors, these systems require a large volume of media, which can be very costly if many recombinant growth factors and small molecules are required. An alternative is to use a miniaturized device, which requires smaller constructs, less cells, and less media, but can be easily reproduced for large scale studies. **Microfluidic systems**, often called **"liver on a chip" devices**, are rectangular chips that are several centimeters long. A central chamber contains a liver construct, often on a millimeter scale. This construct can be a biochip, consisting of a porous membrane layered with cells (similar in concept to a transwell membrane), or may be organoids/spheroids (Fig. 4) (Bhise et al., 2016; Jang et al., 2019; Rennert et al., 2015). The central chamber is linked to microchannels on either side, connected to an inflow and outflow port that allows a uni-directional flow of media through the device, using a micro-pump. Alternatively, the counter-current flow of bile and blood in a liver lobule may be replicated using two parallel channels pumping different media in opposite directions, with one channel flowing through the top compartment of the central chamber and the other channel flowing through the bottom (Jang et al., 2019).

An advantage of these platforms is the high degree of control that can be exerted over biological processes such as cell patterning and substrate stiffness, and the ability to incorporate physical stimuli that mimics physiological conditions such as shear stress, cyclical pressures and mechanical stretch (Vunjak-Novakovic et al., 2021). These systems also allow close monitoring and detailed data collection, by incorporating sensors, high throughput imaging and screening processes, and allowing sampling of the inflow/outflow perfusates (Li, George, Vernetti, Gough, & Taylor, 2018; Rennert et al., 2015).

Furthermore, microfluidic platforms that connect different organ systems have been developed, to examine their interaction during development and normal homeostasis. Such **multi-organ physiological systems** connect different tissue constructs within isolated chambers, including the liver, together with other organs such as intestine, pancreas, kidney, skin, and heart (Bauer et al., 2017; Chang et al., 2017; Maschmeyer et al., 2015; Soltantabar, Calubaquib, Mostafavi, Ghazavi, & Stefan, 2021). The small size of the system makes this logistically feasible, allowing a "whole body" view that is usually only possible with *in vivo* models. These

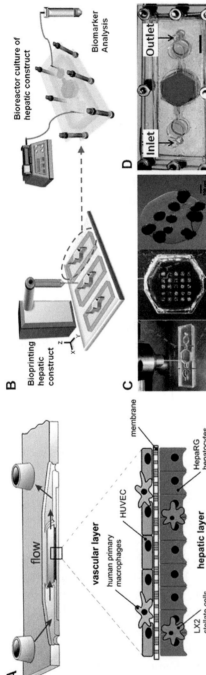

Fig. 4 Microfluidic liver-on-a-chip liver devices, where media flows through a chamber containing 3D liver constructs. (A) A microfluidic device containing a 3D liver biochip, where hepatocytes and stellate cells are cultured on one side of a membrane, and macrophages and endothelial cells (human umbilical vein endothelial cells—HUVEC) are cultured on the opposite side (Rennert et al., 2015). (B) Hepatocyte spheroids are suspended in a GelMA (Gelatin Methacryloyl) hydrogel, and bioprinted as dots to form a hepatocyte construct which is placed in a microfluidic device acting as a bioreactor with media being pumped through the device. (C) GelMA dots containing hepatocyte spheroids are bioprinted in an array format, with each dot containing multiple spheroids. (D) A media-filled device showing the central chamber containing the hepatocyte spheroids, and the inlet and outlet ports for media exchange, with interconnecting channels. *Panels (B–D) From Bhise, N. S., Manoharan, V., Massa, S., Tamayol, A., Ghaderi, M., Miscuglio, M., et al. (2016). A liver-on-a-chip platform with bioprinted hepatic spheroids. Biofabrication, 8(1), 014101. doi:10.1088/1758-5090/8/1/014101.*

multi-organ systems no longer just focus on the local liver niche, but how extra-hepatic niches and systemic signals contribute to liver (and other organ) function, and promise to bridge *in vitro* systems even closer toward the physiological complexity of *in vivo* models.

6. *In vitro* applications of bio-engineered liver platforms
6.1 Developmental biology and normal physiology

There are still many questions about liver development and regeneration that remain unanswered. For example, the exact origin and phenotype of liver progenitor cells in the adult human liver is a matter of debate. However, recent studies have been able to isolate cells with a progenitor-like phenotype from adult human liver, which can form both hepatocyte and cholangiocyte organoids in culture, demonstrating the bipotential plasticity expected of progenitor cells (Huch et al., 2015). The signaling processes and the role of non-parenchymal cells such as stromal cells have been studied in culture to decipher the mechanisms that underlie the progenitor cell response in the human liver (Cordero-Espinoza et al., 2021). Similarly, hiPSC-derived liver organoids have been used to study liver development and identify the interactions between different cell types during this process. hiPSC-derived cells and organoids are less mature than their adult human tissue-derived counterparts, and are usually closer to foetal tissues and cells (Baxter et al., 2015; Ng et al., 2018), and therefore provide a more ethically acceptable and logistically accessible model of early human development without the need for human embryos or foetal tissue. Further questions that may be answered using organoid models include the ontogeny and tissue specification processes that result in liver-specific cells such as liver sinusoidal endothelial cells, Kupffer cells, and hepatic stellate cells. Such models will also facilitate the tracing of how these cells are replaced during liver regeneration, and their role in liver disease. Another major frontier is the use of hiPSC-derived foetal models of organogenesis to study the interaction between different organs as they derive from the germ layers, such as the interaction between the cardiac mesoderm and primitive liver derived from the foregut (Drakhlis et al., 2021), or the development of the liver, pancreas, biliary tree and gallbladder using hiPSC-derived hepato-biliary-pancreatic organoids (Koike et al., 2019).

6.2 Disease modeling and target discovery

Liver injury can arise from a variety of causes, including viral hepatitis, fatty liver disease, excessive alcohol or xenobiotics. Many of these causes share a similar pathophysiological mechanism, with significant alterations to the liver microenvironment. The liver mounts a coordinated response to hepatocyte injury and death, beginning with inflammatory changes including leukocyte recruitment, dedifferentiation of the liver's specialized sinusoidal endothelial cells, activation of hepatic stellate cells, and ECM remodeling (Klaas et al., 2016; Su et al., 2021; Zigmond et al., 2014). Activated stellate cells rapidly transition from quiescent vitamin A and lipid-storing cells into myofibroblasts, and start secreting robust volumes of extracellular matrix, particularly collagen I (Ding et al., 2014; Ramachandran et al., 2019). When the injury is mild and transient in nature, the native hepatocytes proliferate to restore lost parenchyma in a process coordinated by the regenerating liver vasculature, and the liver reverts back to its normal homeostatic state (Ding et al., 2014, 2010; Wei et al., 2021). However, more often liver injury is chronic, and pathophysiological mechanisms such as senescence impairs hepatocyte-based regeneration (Deng et al., 2018; Marshall et al., 2005). In such cases, the liver's pool of facultative progenitor cells, considered to be a subset of the biliary epithelium (cholangiocytes), play a major role in regeneration. Liver progenitor cells accompany an ECM-remodeling process, and proliferate from the portal tracts (where the bile ducts lie) into the parenchyma, within a laminin ECM-rich environment (Kallis et al., 2011; Lorenzini et al., 2010; Raven et al., 2017). Subsequently, they differentiate into hepatocytes to replace lost parenchyma. In many liver diseases the injury is repetitive and not fully resolved by the regenerative response, and an ongoing process of inflammation, progressive fibrosis, and parenchymal destruction occurs (Hoang et al., 2019; Ramachandran et al., 2019). Ultimately this leads to clinical liver failure, as the liver's functions become severely compromised. This inflammatory process is also carcinogenic, and chronic liver disease usually precedes hepatocellular carcinoma in adult patients (Llovet et al., 2021).

Much of our understanding of liver disease and regeneration have traditionally derived from rodent models, which are easy to access, manipulate, and relatively cost-efficient. Rodent findings have often been correlated with human samples, but a major limitation has been that this approach relies on patching together evidence taken from multiple patients at different stages of disease to derive a somewhat artificial narrative. Bio-engineered liver platforms fill in this major gap by allowing experiments that trace the natural progression of disease, by providing reproducible models of human disease that can be assessed over time and compared between different patients and types of models. They also maximize the maturity of hepatocytes used in these models, so they express important receptors that are required for pathogen entry into hepatocytes to effectively model infections. These include sodium taurocholate co-transporting polypeptide (NTCP), a co-transporter for the Hepatitis B virus (Yan et al., 2012), and CD81 and scavenger receptor class B type I, required for malaria-causing plasmodium sporozoites to enter hepatocytes (Manzoni et al., 2017). Such receptors can typically be lost when hepatocytes are cultured as a monolayer, as they dedifferentiate rapidly (Xia et al., 2017). To address this, 3D cultures and co-culture systems have been used to maintain hepatocytes and to also enhance the maturity of stem/progenitor cell sources of hepatocytes to enable the expression of these receptors to study infections (Fu et al., 2019; March et al., 2015).

Major liver conditions such as non-alcoholic fatty liver disease (NAFLD) or hepatocellular carcinoma are complex diseases where single-cell type cultures are unable to recapitulate the disease phenotype *in vitro*. For example, in NAFLD, hyperlipidemia, insulin resistance and oxidative stress not only results in lipid accumulation and toxicity in hepatocytes, but also causes endothelial dysfunction and injury which triggers the activation of hepatic stellate cells and Kupffer cells, with subsequent inflammatory infiltration and fibrosis over time (Powell, Wong, & Rinella, 2021). Complex co-culture systems containing hepatocytes and combinations of non-parenchymal cells (endothelial, Kupffer and stellate cells) have recently been used to develop NAFLD models. These models co-culture cells as a mixed monolayer, cellular aggregates, or self-assembling organoids, and recreate the NAFLD microenvironment by exposing cells to a combination of free fatty acids, fructose, transforming growth factor beta-1 (TGFβ1) and lipopolysaccharide to induce inflammation (Cho et al., 2021; Freag et al., 2021; Slaughter et al., 2021). As hiPSC methods have advanced, recent models have also used multicellular organoids generated using hiPSC derivatives to study the cellular interactions in culture-induced NAFLD (Ouchi et al., 2019; Ramli et al., 2020). Alternatively, cells taken from the diseased liver of patients with NAFLD have been used to create liver organoids to characterize the functional defects and aberrant signaling pathways associated with NAFLD (McCarron et al., 2021), as well as their genetic predisposition to NAFLD (Graffmann et al., 2021).

Liver cancer is another disease that relies on interactions between various cell types, including tumor cells, stromal cells, vasculature, and immune cells. Organoids generated from hepatocellular carcinoma (involving hepatocytes) or cholangiocarcinoma (involving cholangiocytes) contain a heterogenous mixture of cells reflecting the tumor biology, and have been shown to recapitulate the native tumor's histological and molecular phenotype (Broutier et al., 2017; Liu et al., 2021; Saito et al., 2019). These patient-derived cancer organoids have been used to identify potential disease biomarkers and therapeutic avenues, and to investigate the mechanisms of treatment resistance. Personalized cancer organoids are also increasingly being used as a predictive tool to identify patient-specific therapeutic targets and to select drugs with maximum effectiveness and minimal toxicity (Lo, Karlsson, & Kuo, 2020).

6.3 Drug testing

The liver is of major interest in drug development and testing, because it is the primary site for the metabolism of most drugs. Preclinical studies for drug testing allow the assessment and optimization of important parameters of drug metabolism such as pharmacodynamics, pharmacokinetics, and dose-dependent toxicity, as well as the potential for drug-induced liver injury. The battery of tests required for regulatory approval of new drugs includes cell culture studies, as well as validation in animal models to study *in vivo* responses (Ware & Khetani, 2017). Progression of a drug from initial development into clinical trials is highly dependent on accurate preclinical data, to exclude false leads and candidates that are likely to be harmful in patients.

Traditional models of cell culture include primary human hepatocytes in a monolayer or hydrogel-embedded format (usually Matrigel or a collagen sandwich) and are still considered the gold standard in pharmacology (Fraczek, Bolleyn, Vanhaecke, Rogiers, & Vinken, 2013). More recently, immortalized hepatocyte cell lines have provided a more long-lasting, cheaper alternative to primary hepatocytes. However, cell lines generally have a marked reduction in drug transporter and metabolizing enzymes in comparison to primary hepatocytes (Sison-Young et al., 2015), therefore limiting their value. Additionally, there are only a handful of human cell lines, indicating a lack of genetic diversity that makes it difficult to assume that results apply to the highly diverse general population. hiPSC-derived hepatocytes are increasingly becoming a major focus in drug studies because they offer a cheaper, large-scale alternative to primary hepatocytes while accounting for genetic diversity. However, their widespread adoption is currently limited by the lack of standardized protocols that can reliably produce hiPSC-hepatocytes mature enough to recapitulate the full metabolic capacity of hepatocytes (Baxter et al., 2015; Kleiman & Engle, 2021).

Spheroids, organoids, bio-printing, co-cultures with non-parenchymal cells, perfusion bioreactors and microfluidic chips are all strategies used in drug research to maximize the functionality of bioengineered systems (Underhill & Khetani, 2018; Ware & Khetani, 2017), thereby increasing the accuracy of their readouts and their predictive value. These methods not only maintain the metabolic capacity of primary hepatocytes for longer time-points in culture, but also aid in the functional differentiation of hiPSC-hepatocytes, potentially increasing their value in drug research. Multi-organ systems are also particularly useful in this context, because they not only provide the extra-hepatic cues required to stimulate hepatocyte maturation and function, but allow the evaluation of multi-organ interactions in drug metabolism. For example, the metabolism of doxorubicin, a commonly used chemotherapeutic drug with well-known cardiotoxicity, can be investigated using a combined heart/liver microchip. This concurrently facilitates the assessment of doxorubicin metabolism by liver cells, and the cardiotoxic effects of both doxorubicin and its metabolites (produced by the liver cells) on cardiac cells (Soltantabar et al., 2021).

7. Future directions

7.1 Additional cells and structures

While many advances have been made, there are still certain cell types and tissue structures that have not been addressed by the field. This is partly because these aspects are less recognized in liver biology, and techniques to isolate, culture, and assemble these cell types are still not available.

The liver is a major immunological organ, being the first port where blood from the intestine brings food-derived antigens, gut-derived endotoxins, and pathogens. While it contains a large population of liver-resident macrophages (Kupffer cells), it also contains a high density of other myeloid and lymphoid cells, including neutrophils, non-Kupffer macrophages, natural killer cells, and T and B lymphocytes (Gola et al., 2021; Heymann & Tacke, 2016). This immune landscape has not been recreated in any bioengineered liver platforms. Complementary to the immune cells is the lymphatic vasculature, which is still an understudied area in liver biology, but is important to recreate to model fluid and immune regulation in the liver (Tanaka & Iwakiri, 2016). Another important component that also needs consideration is autonomic innervation, which regulates metabolism, blood flow, and bile secretion, as well as the activation of hepatic stellate cells and the progression of liver regeneration (Mizuno & Ueno, 2017).

7.2 Physiological cues

The liver is not an organ that exists in isolation, and in fact interacts with many other organs for example in the regulation of glucose and nutrient metabolism, immune response, hemodynamic control and bile flow. This results in many extra-hepatic cells and signals that influence the liver microenvironment, such as bone marrow-derived cells that travel in blood and are involved in liver regeneration (Almeida-Porada, Zanjani, & Porada, 2010; Wang et al., 2012), and insulin/glucagon signaling from the pancreas that regulates energy metabolism and liver zonation (Cheng et al., 2018). Other physiological cues include the diurnal regulation of liver metabolism and liver size (Sinturel et al., 2017; Wang et al., 2018), and mechanobiology changes that are associated with liver development and disease (Kang, 2020). Many of these physiological factors are not necessarily well understood, and very difficult to recreate *in vitro*. However, bioartificial technologies such as the use of optogenetics (using light to stimulate the expression of specific genes and pathways) (Manoilov, Verkhusha, & Shcherbakova, 2021), electrical stimulation (Chen, Pasricha, Yin, Lin, & Chen, 2010), the delivery of mechanical stimuli such as compression, stretch, and flow using microfluidic chips (Kaarj & Yoon, 2019), as well as multi-organ physiological systems (Maschmeyer et al., 2015) may help in recreating some of these complex physiological cues.

7.3 Designer matrices

While Matrigel has been overwhelmingly popular in bioengineered liver platforms, there is a shift underway to use more tissue-specific or chemically defined matrices, that are free from cancer or animal origins. These studies have shown organoids and liver constructs can be readily derived from non-Matrigel alternatives (Lewis et al., 2018; Saheli et al., 2018; Sorrentino et al., 2020; Ye et al., 2020). However, their widespread adoption is still limited by the lack of a well-characterized commercially available matrix. This is likely to change as commercial interest increases in the clinical applications of bioengineered liver platforms, which place heavy emphasis on xeno-free and standardized reagents. The next generation of chemically defined "designer matrices" will have properties tailored to the cell type being cultured, and will be controlled in a spatio-temporal manner to match different stages of stem cell expansion, self-organization, and differentiation, by using photo-tunable hydrogels, for example (Gjorevski et al., 2016). Effective implementation of such strategies will also depend on greater understanding of the mechanobiological properties of different cell niches within the liver at various stages of development and regeneration.

7.4 Liver zonation

Zonation is an essential structural and functional organization in the liver. Although its underlying mechanism is still under investigation, major influences have been uncovered, such as oxygen and nutrient gradients, regulatory hormones, and Wnt signaling (Gebhardt & Matz-Soja, 2014). The functional zonation of hepatocytes *in vitro* has been achieved by modulating some of these factors, such as creating an oxygen gradient (Janani & Mandal, 2021; Tonon et al., 2019), and controlling the expression of β-catenin to create a Wnt signaling gradient (Wahlicht et al., 2020). These strategies should be applied to multi-cellular liver constructs to not only recapitulate the zonation of non-hepatocyte cell types, but to model the basis and effects of zonation on liver tissue as a whole, factoring in the interactions between different cell types.

8. Conclusion

The liver is a structurally and functionally complex organ, and its unique microenvironment is critical for this. Methods to recreate this niche requires careful consideration of the cell composition, source of cells, media ingredients, extracellular matrix, and bio-engineered technologies such as scaffolds, 3D bio-printing, perfusion bioreactors, and microfluidic systems. The field has transitioned from being reliant on rodent cells, to a focus on generating human liver constructs using primary human liver cells and hiPSC-derived cells, increasing the clinical relevance of these products. More complex models with greater fidelity to native liver tissue and its microenvironment is expected in the future, by incorporating further cell types and structures, and using strategies such as designer matrices. These bio-engineered liver platforms are useful for studying liver development and biology, and can also be used to model disease conditions, identify new therapeutic targets, and play a major role in drug development.

Acknowledgments

The authors gratefully acknowledge funding from the Australian National Health & Medical Research Council (NHMRC), Stafford Fox Medical Research Foundation, O'Brien Institute Foundation, St Vincent's Institute Foundation, St Vincent's Hospital Melbourne Research Endowment Fund, Australian Catholic University, University of Melbourne Centre for Stem Cell Systems, Bioplatforms Australia, Australian and New Zealand Hepatic, Pancreatic and Biliary Association, and the Victorian State Government's Department of Business Innovation Operational Infrastructure Support Program.

Conflict of interest

None to declare.

References

Aisenbrey, E. A., & Murphy, W. L. (2020). Synthetic alternatives to Matrigel. *Nature Reviews Materials, 5*(7), 539–551. https://doi.org/10.1038/s41578-020-0199-8.

Aizarani, N., Saviano, A., Sagar, Mailly, L., Durand, S., Herman, J. S., et al. (2019). A human liver cell atlas reveals heterogeneity and epithelial progenitors. *Nature, 572*(7768), 199–204. https://doi.org/10.1038/s41586-019-1373-2.

Almeida-Porada, G., Zanjani, E. D., & Porada, C. D. (2010). Bone marrow stem cells and liver regeneration. *Experimental Hematology, 38*(7), 574–580. https://doi.org/10.1016/j.exphem.2010.04.007.

Ang, L. T., Tan, A. K. Y., Autio, M. I., Goh, S. H., Choo, S. H., Lee, K. L., et al. (2018). A roadmap for human liver differentiation from pluripotent stem cells. *Cell Reports, 22*(8), 2190–2205. https://doi.org/10.1016/j.celrep.2018.01.087.

Arai, T., Sakurai, T., Kamiyoshi, A., Ichikawa-Shindo, Y., Iinuma, N., Iesato, Y., et al. (2011). Induction of LYVE-1/stabilin-2-positive liver sinusoidal endothelial-like cells from embryoid bodies by modulation of adrenomedullin-RAMP2 signaling. *Peptides, 32*(9), 1855–1865. https://doi.org/10.1016/j.peptides.2011.07.005.

Arteel, G. E., & Naba, A. (2020). The liver matrisome—Looking beyond collagens. *JHEP Reports, 2*(4). https://doi.org/10.1016/j.jhep.2020.100115, 100115.

Auth, M. K., Joplin, R. E., Okamoto, M., Ishida, Y., McMaster, P., Neuberger, J. M., et al. (2001). Morphogenesis of primary human biliary epithelial cells: Induction in high-density culture or by coculture with autologous human hepatocytes. *Hepatology, 33*(3), 519–529. https://doi.org/10.1053/jhep.2001.22703.

Bachour-El Azzi, P., Sharanek, A., Burban, A., Li, R., Guével, R. L., Abdel-Razzak, Z., et al. (2015). Comparative localization and functional activity of the main hepatobiliary transporters in HepaRG cells and primary human hepatocytes. *Toxicological Sciences, 145*(1), 157–168. https://doi.org/10.1093/toxsci/kfv041.

Badylak, S. F., Taylor, D., & Uygun, K. (2011). Whole-organ tissue engineering: Decellularization and recellularization of three-dimensional matrix scaffolds. *Annual Review of Biomedical Engineering, 13*, 27–53. https://doi.org/10.1146/annurev-bioeng-071910-124743.

Bale, S. S., Golberg, I., Jindal, R., McCarty, W. J., Luitje, M., Hegde, M., et al. (2015). Long-term coculture strategies for primary hepatocytes and liver sinusoidal endothelial cells. *Tissue Engineering. Part C, Methods, 21*(4), 413–422. https://doi.org/10.1089/ten.TEC.2014.0152.

Banales, J. M., Huebert, R. C., Karlsen, T., Strazzabosco, M., LaRusso, N. F., & Gores, G. J. (2019). Cholangiocyte pathobiology. *Nature Reviews. Gastroenterology & Hepatology, 16*(5), 269–281. https://doi.org/10.1038/s41575-019-0125-y.

Banas, A., Teratani, T., Yamamoto, Y., Tokuhara, M., Takeshita, F., & Quinn, G. (2007). Adipose tissue-derived mesenchymal stem cells as a source of human hepatocytes. *Hepatology*, *46*(1), 219–228. https://doi.org/10.1002/hep.21704.

Baptista, P. M., Siddiqui, M. M., Lozier, G., Rodriguez, S. R., Atala, A., & Soker, S. (2011). The use of whole organ decellularization for the generation of a vascularized liver organoid. *Hepatology*, *53*(2), 604–617. https://doi.org/10.1002/hep.24067.

Bauer, S., Wennberg Huldt, C., Kanebratt, K. P., Durieux, I., Gunne, D., Andersson, S., et al. (2017). Functional coupling of human pancreatic islets and liver spheroids on-a-chip: Towards a novel human ex vivo type 2 diabetes model. *Scientific Reports*, *7*(1), 14620. https://doi.org/10.1038/s41598-017-14815-w.

Baxter, M., Withey, S., Harrison, S., Segeritz, C.-P., Zhang, F., Atkinson-Dell, R., et al. (2015). Phenotypic and functional analyses show stem cell-derived hepatocyte-like cells better mimic fetal rather than adult hepatocytes. *Journal of Hepatology*, *62*(3), 581–589. https://doi.org/10.1016/j.jhep.2014.10.016.

Bell, C. C., Hendriks, D. F. G., Moro, S. M. L., Ellis, E., Walsh, J., Renblom, A., et al. (2016). Characterization of primary human hepatocyte spheroids as a model system for drug-induced liver injury, liver function and disease. *Scientific Reports*, *6*, 25187. https://doi.org/10.1038/srep25187.

Ben-Moshe, S., & Itzkovitz, S. (2019). Spatial heterogeneity in the mammalian liver. *Nature Reviews Gastroenterology & Hepatology*, *16*(7), 395–410. https://doi.org/10.1038/s41575-019-0134-x.

Ben-Moshe, S., Shapira, Y., Moor, A. E., Manco, R., Veg, T., Bahar Halpern, K., et al. (2019). Spatial sorting enables comprehensive characterization of liver zonation. *Nature Metabolism*, *1*(9), 899–911. https://doi.org/10.1038/s42255-019-0109-9.

Berardis, S., Lombard, C., Evraerts, J., El Taghdouini, A., Rosseels, V., Sancho-Bru, P., et al. (2014). Gene expression profiling and secretome analysis differentiate adult-derived human liver stem/progenitor cells and human hepatic stellate cells. *PLoS One*, *9*(1). https://doi.org/10.1371/journal.pone.0086137, e86137.

Bhandari, S., Li, R., Simón-Santamaría, J., McCourt, P., Johansen, S. D., Smedsrød, B., et al. (2020). Transcriptome and proteome profiling reveal complementary scavenger and immune features of rat liver sinusoidal endothelial cells and liver macrophages. *BMC Molecular and Cell Biology*, *21*(1), 85. https://doi.org/10.1186/s12860-020-00331-9.

Bhise, N. S., Manoharan, V., Massa, S., Tamayol, A., Ghaderi, M., Miscuglio, M., et al. (2016). A liver-on-a-chip platform with bioprinted hepatic spheroids. *Biofabrication*, *8*(1). https://doi.org/10.1088/1758-5090/8/1/014101, 014101.

Bhogal, R. H., Hodson, J., Bartlett, D. C., Weston, C. J., Curbishley, S. M., Haughton, E., et al. (2011). Isolation of primary human hepatocytes from normal and diseased liver tissue: A one hundred liver experience. *PLoS One*, *6*(3). https://doi.org/10.1371/journal.pone.0018222, e18222.

Bianconi, E., Piovesan, A., Facchin, F., Beraudi, A., Casadei, R., Frabetti, F., et al. (2013). An estimation of the number of cells in the human body. *Annals of Human Biology*, *40*(6), 463–471. https://doi.org/10.3109/03014460.2013.807878.

Bishi, D. K., Mathapati, S., Venugopal, J. R., Guhathakurta, S., Cherian, K. M., Verma, R. S., et al. (2016). A patient-inspired ex vivo liver tissue engineering approach with autologous mesenchymal stem cells and hepatogenic serum. *Advanced Healthcare Materials*, *5*(9), 1058–1070. https://doi.org/10.1002/adhm.201500897.

Blaeser, A., Duarte Campos, D., Puster, U., Richtering, W., Stevens, M., & Fischer, H. (2015). Controlling shear stress in 3D bioprinting is a key factor to balance printing resolution and stem cell integrity. *Advanced Healthcare Materials*, *5*. https://doi.org/10.1002/adhm.201500677.

Blouin, A., Bolender, R. P., & Weibel, E. R. (1977). Distribution of organelles and membranes between hepatocytes and nonhepatocytes in the rat liver parenchyma. A stereological study. *Journal of Cell Biology, 72*(2), 441–455. https://doi.org/10.1083/jcb.72.2.441.

Boyer, J. L., & Soroka, C. J. (2021). Bile formation and secretion: An update. *Journal of Hepatology, 75*(1), 190–201. https://doi.org/10.1016/j.jhep.2021.02.011.

Brophy, C. M., Luebke-Wheeler, J. L., Amiot, B. P., Khan, H., Remmel, R. P., Rinaldo, P., et al. (2009). Rat hepatocyte spheroids formed by rocked technique maintain differentiated hepatocyte gene expression and function. *Hepatology, 49*(2), 578–586. https://doi.org/10.1002/hep.22674.

Brosch, M., Kattler, K., Herrmann, A., von Schonfels, W., Nordstrom, K., Seehofer, D., et al. (2018). Epigenomic map of human liver reveals principles of zonated morphogenic and metabolic control. *Nature Communications, 9*(1), 4150. https://doi.org/10.1038/s41467-018-06611-5.

Broutier, L., Mastrogiovanni, G., Verstegen, M. M., Francies, H. E., Gavarro, L. M., Bradshaw, C. R., et al. (2017). Human primary liver cancer-derived organoid cultures for disease modeling and drug screening. *Nature Medicine, 23*(12), 1424–1435. https://doi.org/10.1038/nm.4438.

Brown, J. H., Das, P., DiVito, M. D., Ivancic, D., Tan, L. P., & Wertheim, J. A. (2018). Nanofibrous PLGA electrospun scaffolds modified with type I collagen influence hepatocyte function and support viability in vitro. *Acta Biomaterialia, 73*, 217–227. https://doi.org/10.1016/j.actbio.2018.02.009.

Caliari, S. R., & Burdick, J. A. (2016). A practical guide to hydrogels for cell culture. *Nature Methods, 13*(5), 405–414. https://doi.org/10.1038/nmeth.3839.

Campana, L., Esser, H., Huch, M., & Forbes, S. (2021). Liver regeneration and inflammation: From fundamental science to clinical applications. *Nature Reviews Molecular Cell Biology, 22*(9), 608–624. https://doi.org/10.1038/s41580-021-00373-7.

Cao, Y., Mitchell, G., Messina, A., Price, L., Thompson, E., Penington, A., et al. (2006). The influence of architecture on degradation and tissue ingrowth into three-dimensional poly(lactic-co-glycolic acid) scaffolds in vitro and in vivo. *Biomaterials, 27*(14), 2854–2864. https://doi.org/10.1016/j.biomaterials.2005.12.015.

Chan, S. C., Liu, C. L., Lo, C. M., Lam, B. K., Lee, E. W., Wong, Y., et al. (2006). Estimating liver weight of adults by body weight and gender. *World Journal of Gastroenterology, 12*(14), 2217–2222. https://doi.org/10.3748/wjg.v12.i4.2217.

Chang, S. Y., Weber, E. J., Sidorenko, V. S., Chapron, A., Yeung, C. K., Gao, C., et al. (2017). Human liver-kidney model elucidates the mechanisms of aristolochic acid nephrotoxicity. *JCI Insight, 2*(22). https://doi.org/10.1172/jci.insight.95978.

Chen, J., Pasricha, P. J., Yin, J., Lin, L., & Chen, J. D. Z. (2010). Hepatic electrical stimulation reduces blood glucose in diabetic rats. *Neurogastroenterology and Motility, 22*(10), 1109–e1286. https://doi.org/10.1111/j.1365-2982.2010.01556.x.

Chen, F., Wang, H., & Xiao, J. (2020). Regulated differentiation of stem cells into an artificial 3D liver as a transplantable source. *Clinical and Molecular Hepatology, 26*(2), 163–179. https://doi.org/10.3350/cmh.2019.0022n.

Cheng, X., Kim, S. Y., Okamoto, H., Xin, Y., Yancopoulos, G. D., Murphy, A. J., et al. (2018). Glucagon contributes to liver zonation. *Proceedings of the National Academy of Sciences of the United States of America, 115*(17), E4111–E4119. https://doi.org/10.1073/pnas.1721403115.

Chitrangi, S., Nair, P., & Khanna, A. (2017). Three-dimensional polymer scaffolds for enhanced differentiation of human mesenchymal stem cells to hepatocyte-like cells: A comparative study. *Journal of Tissue Engineering and Regenerative Medicine, 11*(8), 2359–2372. https://doi.org/10.1002/term.2136.

Cho, H.-J., Kim, H.-J., Lee, K., Lasli, S., Ung, A., Hoffman, T., et al. (2021). Bioengineered multicellular liver microtissues for modeling advanced hepatic fibrosis driven through non-alcoholic fatty liver disease. *Small, 17*(14), 2007425. https://doi.org/10.1002/smll.202007425.

Christoffersson, J., Aronsson, C., Jury, M., Selegard, R., Aili, D., & Mandenius, C. F. (2018). Fabrication of modular hyaluronan-PEG hydrogels to support 3D cultures of hepatocytes in a perfused liver-on-a-chip device. *Biofabrication, 11*(1). https://doi.org/10.1088/1758-5090/aaf657, 015013.

Coger, R., Toner, M., Moghe, P., Ezzell, R. M., & Yarmush, M. L. (1997). Hepatocyte aggregation and reorganization of EHS matrix gel. *Tissue Engineering, 3*(4), 375–390. https://doi.org/10.1089/ten.1997.3.375.

Coll, M., Perea, L., Boon, R., Leite, S. B., Vallverdu, J., Mannaerts, I., et al. (2018). Generation of hepatic stellate cells from human pluripotent stem cells enables in vitro modeling of liver fibrosis. *Cell Stem Cell, 23*(1), 101–113. e107 https://doi.org/10.1016/j.stem.2018.05.027.

Cordero-Espinoza, L., Dowbaj, A. M., Kohler, T. N., Strauss, B., Sarlidou, O., Belenguer, G., et al. (2021). Dynamic cell contacts between periportal mesenchyme and ductal epithelium act as a rheostat for liver cell proliferation. *Cell Stem Cell.* https://doi.org/10.1016/j.stem.2021.07.002 (Epublication ahead of print, 2 August 2021).

Coronado, R. E., Somaraki-Cormier, M., Natesan, S., Christy, R. J., Ong, J. L., & Halff, G. A. (2017). Decellularization and solubilization of porcine liver for use as a substrate for porcine hepatocyte culture: Method optimization and comparison. *Cell Transplantation, 26*(12), 1840–1854. https://doi.org/10.1177/0963689717742157.

Couvelard, A., Bringuier, A. F., Dauge, M. C., Nejjari, M., Darai, E., Benifla, J. L., et al. (1998). Expression of integrins during liver organogenesis in humans. *Hepatology, 27*(3), 839–847. S0270913998001220 [pii].

Croce, S., Peloso, A., Zoro, T., Avanzini, M. A., & Cobianchi, L. (2019). A hepatic scaffold from decellularized liver tissue: Food for thought. *Biomolecules, 9*(12), 813. https://doi.org/10.3390/biom9120813.

Cuvellier, M., Ezan, F., Oliveira, H., Rose, S., Fricain, J.-C., Langouët, S., et al. (2021). 3D culture of HepaRG cells in GelMa and its application to bioprinting of a multicellular hepatic model. *Biomaterials, 269.* https://doi.org/10.1016/j.biomaterials.2020.120611, 120611.

Daneshgar, A., Klein, O., Nebrich, G., Weinhart, M., Tang, P., Arnold, A., et al. (2020). The human liver matrisome—Proteomic analysis of native and fibrotic human liver extracellular matrices for organ engineering approaches. *Biomaterials, 257.* https://doi.org/10.1016/j.biomaterials.2020.120247, 120247.

Danoy, M., Tauran, Y., Poulain, S., Jellali, R., Bruce, J., Leduc, M., et al. (2021). Investigation of the hepatic development in the coculture of hiPSCs-derived LSECs and HLCs in a fluidic microenvironment. *APL Bioengineering, 5*(2). https://doi.org/10.1063/5.0041227, 026104.

Dao Thi, V. L., Wu, X., Belote, R. L., Andreo, U., Takacs, C. N., Fernandez, J. P., et al. (2020). Stem cell-derived polarized hepatocytes. *Nature Communications, 11*(1), 1677. https://doi.org/10.1038/s41467-020-15337-2.

De Smedt, J., van Os, E. A., Talon, I., Ghosh, S., Toprakhisar, B., Da Costa Furtado Madeiro, R., et al. (2021). PU.1 drives specification of pluripotent stem cell-derived endothelial cells to LSEC-like cells. *Cell Death & Disease, 12*(1), 84. https://doi.org/10.1038/s41419-020-03356-2.

Deegan, D. B., Zimmerman, C., Skardal, A., Atala, A., & Shupe, T. D. (2015). Stiffness of hyaluronic acid gels containing liver extracellular matrix supports human hepatocyte function and alters cell morphology. *Journal of the Mechanical Behavior of Biomedical Materials, 55*, 87–103. https://doi.org/10.1016/j.jmbbm.2015.10.016.

DeLeve, L. D., Wang, X., & Guo, Y. (2008). Sinusoidal endothelial cells prevent rat stellate cell activation and promote reversion to quiescence. *Hepatology*, *48*(3), 920–930. https://doi.org/10.1002/hep.22351.

DeLeve, L. D., Wang, X., Hu, L., McCuskey, M. K., & McCuskey, R. S. (2004). Rat liver sinusoidal endothelial cell phenotype is maintained by paracrine and autocrine regulation. *American Journal of Physiology. Gastrointestinal and Liver Physiology*, *287*(4), G757–G763. https://doi.org/10.1152/ajpgi.00017.2004.

Deng, X., Zhang, X., Li, W., Feng, R.-X., Li, L., Yi, G.-R., et al. (2018). Chronic liver injury induces conversion of biliary epithelial cells into hepatocytes. *Cell Stem Cell*, *23*(1), 114–122.e113. https://doi.org/10.1016/j.stem.2018.05.022.

Dianat, N., Dubois-Pot-Schneider, H., Steichen, C., Desterke, C., Leclerc, P., Raveux, A., et al. (2014). Generation of functional cholangiocyte-like cells from human pluripotent stem cells and HepaRG cells. *Hepatology*, *60*(2), 700–714. https://doi.org/10.1002/hep.27165.

Ding, B. S., Cao, Z., Lis, R., Nolan, D. J., Guo, P., Simons, M., et al. (2014). Divergent angiocrine signals from vascular niche balance liver regeneration and fibrosis. *Nature*, *505*(7481), 97–102. https://doi.org/10.1038/nature12681.

Ding, B. S., Nolan, D. J., Butler, J. M., James, D., Babazadeh, A. O., Rosenwaks, Z., et al. (2010). Inductive angiocrine signals from sinusoidal endothelium are required for liver regeneration. *Nature*, *468*(7321), 310–315. https://doi.org/10.1038/nature09493.

Dingle, A. M., Yap, K. K., Gerrand, Y.-W., Taylor, C. J., Keramidaris, E., Lokmic, Z., et al. (2018). Characterization of isolated liver sinusoidal endothelial cells for liver bioengineering. *Angiogenesis*, *21*(3), 581–597. https://doi.org/10.1007/s10456-018-9610-0.

Dobie, R., Wilson-Kanamori, J. R., Henderson, B. E. P., Smith, J. R., Matchett, K. P., Portman, J. R., et al. (2019). Single-cell transcriptomics uncovers zonation of function in the mesenchyme during liver fibrosis. *Cell Reports*, *29*(7), 1832–1847. e1838 https://doi.org/10.1016/j.celrep.2019.10.024.

Drakhlis, L., Biswanath, S., Farr, C.-M., Lupanow, V., Teske, J., Ritzenhoff, K., et al. (2021). Human heart-forming organoids recapitulate early heart and foregut development. *Nature Biotechnology*, *39*(6), 737–746. https://doi.org/10.1038/s41587-021-00815-9.

Dunn, J. C., Yarmush, M. L., Koebe, H. G., & Tompkins, R. G. (1989). Hepatocyte function and extracellular matrix geometry: Long-term culture in a sandwich configuration. *The FASEB Journal*, *3*(2), 174–177. https://doi.org/10.1096/fasebj.3.2.2914628.

Dvir-Ginzberg, M., Gamlieli-Bonshtein, I., Agbaria, R., & Cohen, S. (2003). Liver tissue engineering within alginate scaffolds: Effects of cell-seeding density on hepatocyte viability, morphology, and function. *Tissue Engineering*, *9*(4), 757–766. https://doi.org/10.1089/107632703768247430.

Enomoto, Y., Enomura, M., Takebe, T., Mitsuhashi, Y., Kimura, M., Yoshizawa, E., et al. (2014). Self-formation of vascularized hepatic tissue from human adult hepatocyte. *Transplantation Proceedings*, *46*(4), 1243–1246. https://doi.org/10.1016/j.transproceed.2013.11.086.

Faulk, D. M., Carruthers, C. A., Warner, H. J., Kramer, C. R., Reing, J. E., Zhang, L., et al. (2014). The effect of detergents on the basement membrane complex of a biologic scaffold material. *Acta Biomaterialia*, *10*(1), 183–193. https://doi.org/10.1016/j.actbio.2013.09.006.

Feng, Z. Q., Chu, X., Huang, N. P., Wang, T., Wang, Y., Shi, X., et al. (2009). The effect of nanofibrous galactosylated chitosan scaffolds on the formation of rat primary hepatocyte aggregates and the maintenance of liver function. *Biomaterials*, *30*(14), 2753–2763. https://doi.org/10.1016/j.biomaterials.2009.01.053.

Fey, S. J., & Wrzesinski, K. (2012). Determination of drug toxicity using 3D spheroids constructed from an immortal human hepatocyte cell line. *Toxicological sciences: an official journal of the Society of Toxicology*, *127*(2), 403–411. https://doi.org/10.1093/toxsci/kfs122.

Fishman, J. M., Lowdell, M. W., Urbani, L., Ansari, T., Burns, A. J., Turmaine, M., et al. (2013). Immunomodulatory effect of a decellularized skeletal muscle scaffold in a discordant xenotransplantation model. *Proceedings of the National Academy of Sciences*, *110*(35), 14360–14365. https://doi.org/10.1073/pnas.1213228110.

Fitzpatrick, E., Wu, Y., Dhadda, P., Hughes, R. D., Mitry, R. R., Qin, H., et al. (2015). Coculture with mesenchymal stem cells results in improved viability and function of human hepatocytes. *Cell Transplantation*, *24*(1), 73–83. https://doi.org/10.3727/096368913X674080.

Fraczek, J., Bolleyn, J., Vanhaecke, T., Rogiers, V., & Vinken, M. (2013). Primary hepatocyte cultures for pharmaco-toxicological studies: At the busy crossroad of various anti-dedifferentiation strategies. *Archives of Toxicology*, *87*(4), 577–610. https://doi.org/10.1007/s00204-012-0983-3.

Freag, M. S., Namgung, B., Reyna Fernandez, M. E., Gherardi, E., Sengupta, S., & Jang, H. L. (2021). Human nonalcoholic steatohepatitis on a chip. *Hepatology Communications*, *5*(2), 217–233. https://doi.org/10.1002/hep4.1647.

Frenkel, N. C., Poghosyan, S., Verheem, A., Padera, T. P., Rinkes, I. H. M. B., Kranenburg, O., et al. (2020). Liver lymphatic drainage patterns follow segmental anatomy in a murine model. *Scientific Reports*, *10*(1), 21808. https://doi.org/10.1038/s41598-020-78727-y.

Fu, G.-B., Huang, W.-J., Zeng, M., Zhou, X., Wu, H.-P., Liu, C.-C., et al. (2019). Expansion and differentiation of human hepatocyte-derived liver progenitor-like cells and their use for the study of hepatotropic pathogens. *Cell Research*, *29*(1), 8–22. https://doi.org/10.1038/s41422-018-0103-x.

Gage, B. K., Liu, J. C., Innes, B. T., MacParland, S. A., McGilvray, I. D., Bader, G. D., et al. (2020). Generation of functional liver sinusoidal endothelial cells from human pluripotent stem-cell-derived venous angioblasts. *Cell Stem Cell*, *27*(2), 254–269 e259. https://doi.org/10.1016/j.stem.2020.06.007.

Gaignerie, A., Lefort, N., Rousselle, M., Forest-Choquet, V., Flippe, L., Francois–Campion, V., et al. (2018). Urine-derived cells provide a readily accessible cell type for feeder-free mRNA reprogramming. Scientific Reports, 8(1), 14363. doi:https://doi.org/10.1038/s41598-018-32645-2.

Gebhardt, R., & Matz-Soja, M. (2014). Liver zonation: Novel aspects of its regulation and its impact on homeostasis. *World Journal of Gastroenterology*, *20*(26), 8491–8504. https://doi.org/10.3748/wjg.v20.i26.8491.

Gissen, P., & Arias, I. M. (2015). Structural and functional hepatocyte polarity and liver disease. *Journal of Hepatology*, *63*(4), 1023–1037. https://doi.org/10.1016/j.jhep.2015.06.015.

Gjorevski, N., Sachs, N., Manfrin, A., Giger, S., Bragina, M. E., Ordóñez-Morán, P., et al. (2016). Designer matrices for intestinal stem cell and organoid culture. *Nature*, *539*(7630), 560–564. https://doi.org/10.1038/nature20168.

Gola, A., Dorrington, M. G., Speranza, E., Sala, C., Shih, R. M., Radtke, A. J., et al. (2021). Commensal-driven immune zonation of the liver promotes host defence. *Nature*, *589*(7840), 131–136. https://doi.org/10.1038/s41586-020-2977-2.

Gomez-Lechon, M. J., Jover, R., Donato, T., Ponsoda, X., Rodriguez, C., Stenzel, K. G., et al. (1998). Long-term expression of differentiated functions in hepatocytes cultured in three-dimensional collagen matrix. *Journal of Cellular Physiology*, *177*(4), 553–562. https://doi.org/10.1002/(SICI)1097-4652(199812)177:4<553::AID-JCP6>3.0.CO;2-F.

Gopalan, N., Nor, S. N. M., & Mohamed, M. S. (2018). Global human embryonic stem cell laws and policies and their influence on stem cell tourism. *Biotechnology Law Report*, *37*(5), 255–269. https://doi.org/10.1089/blr.2018.29088.ng.

Graffmann, N., Ncube, A., Martins, S., Fiszl, A. R., Reuther, P., Bohndorf, M., et al. (2021). A stem cell based in vitro model of NAFLD enables the analysis of patient specific individual metabolic adaptations in response to a high fat diet and AdipoRon interference. *Biology Open*, *10*(1). https://doi.org/10.1242/bio.054189.

Grant, R., Hallett, J., Forbes, S., Hay, D., & Callanan, A. (2019). Blended electrospinning with human liver extracellular matrix for engineering new hepatic microenvironments. *Scientific Reports*, *9*(1), 6293. https://doi.org/10.1038/s41598-019-42627-7.

Grant, R., Hay, D. C., & Callanan, A. (2017). A drug-induced hybrid electrospun poly-capro-lactone: Cell-derived extracellular matrix scaffold for liver tissue engineering. *Tissue Engineering. Part A*, *23*(13-14), 650–662. https://doi.org/10.1089/ten.TEA.2016.0419.

Gratte, F. D., Pasic, S., Olynyk, J. K., Yeoh, G. C. T., Tosh, D., Coombe, D. R., et al. (2018). Transdifferentiation of pancreatic progenitor cells to hepatocyte-like cells is not serum-dependent when facilitated by extracellular matrix proteins. *Scientific Reports*, *8*(1), 4385. https://doi.org/10.1038/s41598-018-22596-z.

Gressner, A. M., Krull, N., & Bachem, M. G. (1994). Regulation of proteoglycan expression in fibrotic liver and cultured fat-storing cells. *Pathology, Research and Practice*, *190*(9-10), 864–882. https://doi.org/10.1016/S0344-0338(11)80990-8.

Gungor-Ozkerim, P. S., Inci, I., Zhang, Y. S., Khademhosseini, A., & Dokmeci, M. R. (2018). Bioinks for 3D bioprinting: An overview. *Biomaterials Science*, *6*(5), 915–946. https://doi.org/10.1039/c7bm00765e.

Hafiz, E. O. A., Bulutoglu, B., Mansy, S. S., Chen, Y., Abu-Taleb, H., Soliman, S. A. M., et al. (2021). Development of liver microtissues with functional biliary ductular network. *Biotechnology and Bioengineering*, *118*(1), 17–29. https://doi.org/10.1002/bit.27546.

Halpern, K. B., Shenhav, R., Massalha, H., Toth, B., Egozi, A., Massasa, E. E., et al. (2018). Paired-cell sequencing enables spatial gene expression mapping of liver endothelial cells. *Nature Biotechnology*, *36*(10), 962–970. https://doi.org/10.1038/nbt.4231.

Halpern, K. B., Shenhav, R., Matcovitch-Natan, O., Toth, B., Lemze, D., Golan, M., et al. (2017). Single-cell spatial reconstruction reveals global division of labour in the mammalian liver. *Nature*, *542*(7641), 352–356. https://doi.org/10.1038/nature21065.

Heslop, J. A., Rowe, C., Walsh, J., Sison-Young, R., Jenkins, R., Kamalian, L., et al. (2017). Mechanistic evaluation of primary human hepatocyte culture using global proteomic analysis reveals a selective dedifferentiation profile. *Archives of Toxicology*, *91*(1), 439–452. https://doi.org/10.1007/s00204-016-1694-y.

Heymann, F., & Tacke, F. (2016). Immunology in the liver—from homeostasis to disease. *Nature Reviews Gastroenterology & Hepatology*, *13*(2), 88–110. https://doi.org/10.1038/nrgastro.2015.200.

Hoang, S. A., Oseini, A., Feaver, R. E., Cole, B. K., Asgharpour, A., Vincent, R., et al. (2019). Gene expression predicts histological severity and reveals distinct molecular profiles of nonalcoholic fatty liver disease. *Scientific Reports*, *9*(1), 12541. https://doi.org/10.1038/s41598-019-48746-5.

Hosseini, V., Maroufi, N. F., Saghati, S., Asadi, N., Darabi, M., Ahmad, S. N. S., et al. (2019). Current progress in hepatic tissue regeneration by tissue engineering. *Journal of Translational Medicine*, *17*(1), 383. https://doi.org/10.1186/s12967-019-02137-6.

Hou, Y. T., & Hsu, C. C. (2020). Development of a 3D porous chitosan/gelatin liver scaffold for a bioartificial liver device. *Journal of Bioscience and Bioengineering*, *129*(6), 741–748. https://doi.org/10.1016/j.jbiosc.2019.12.012.

Hu, H., Gehart, H., Artegiani, B., López-Iglesias, C., Dekkers, F., Basak, O., et al. (2018). Long-term expansion of functional mouse and human hepatocytes as 3D organoids. *Cell*, *175*(6), 1591–1606. e1519 https://doi.org/10.1016/j.cell.2018.11.013.

Huch, M., Dorrell, C., Boj, S. F., van Es, J. H., Li, V. S., van de Wetering, M., et al. (2013). In vitro expansion of single Lgr5 + liver stem cells induced by Wnt-driven regeneration. *Nature*, *494*(7436), 247–250. https://doi.org/10.1038/nature11826.

Huch, M., Gehart, H., van Boxtel, R., Hamer, K., Blokzijl, F., Verstegen, M. M., et al. (2015). Long-term culture of genome-stable bipotent stem cells from adult human liver. *Cell*, *160*(1-2), 299–312. https://doi.org/10.1016/j.cell.2014.11.050.

Hughes, C. S., Postovit, L. M., & Lajoie, G. A. (2010). Matrigel: A complex protein mixture required for optimal growth of cell culture. *Proteomics*, *10*(9), 1886–1890. https://doi.org/10.1002/pmic.200900758.

Hurrell, T., Ellero, A. A., Masso, Z. F., & Cromarty, A. D. (2018). Characterization and reproducibility of HepG2 hanging drop spheroids toxicology in vitro. *Toxicology In Vitro*, *50*, 86–94. https://doi.org/10.1016/j.tiv.2018.02.013.

Hussein, K. H., Park, K. M., Yu, L., Kwak, H. H., & Woo, H. M. (2020). Decellularized hepatic extracellular matrix hydrogel attenuates hepatic stellate cell activation and liver fibrosis. *Materials Science & Engineering. C, Materials for Biological Applications*, *116*. https://doi.org/10.1016/j.msec.2020.111160, 111160.

Hussey, G. S., Dziki, J. L., & Badylak, S. F. (2018). Extracellular matrix-based materials for regenerative medicine. *Nature Reviews Materials*, *3*(7), 159–173. https://doi.org/10.1038/s41578-018-0023-x.

Ijima, H., Hou, Y.-T., & Takei, T. (2010). Development of hepatocyte-embedded hydrogel-filled macroporous scaffold cultures using transglutaminase. *Biochemical Engineering Journal*, *52*(2), 276–281. https://doi.org/10.1016/j.bej.2010.09.003.

Inamori, M., Mizumoto, H., & Kajiwara, T. (2009). An approach for formation of vascularized liver tissue by endothelial cell-covered hepatocyte spheroid integration. *Tissue Engineering. Part A*, *15*(8), 2029–2037. https://doi.org/10.1089/ten.tea.2008.0403.

Janani, G., & Mandal, B. B. (2021). Mimicking physiologically relevant hepatocyte zonation using immunomodulatory silk liver extracellular matrix scaffolds toward a bioartificial liver platform. *ACS Applied Materials & Interfaces*, *13*(21), 24401–24421. https://doi.org/10.1021/acsami.1c00719.

Jang, K. J., Otieno, M. A., Ronxhi, J., Lim, H. K., Ewart, L., Kodella, K. R., et al. (2019). Reproducing human and cross-species drug toxicities using a liver-chip. *Science Translational Medicine*, *11*(517). https://doi.org/10.1126/scitranslmed.aax5516.

Juza, R. M., & Pauli, E. M. (2014). Clinical and surgical anatomy of the liver: A review for clinicians. *Clinical Anatomy*, *27*(5), 764–769. https://doi.org/10.1002/ca.22350.

Kaarj, K., & Yoon, J. Y. (2019). Methods of delivering mechanical stimuli to organ-on-a-chip. *Micromachines (Basel)*, *10*(10). https://doi.org/10.3390/mi10100700.

Kallis, Y. N., Robson, A. J., Fallowfield, J. A., Thomas, H. C., Alison, M. R., Wright, N. A., et al. (2011). Remodelling of extracellular matrix is a requirement for the hepatic progenitor cell response. *Gut*, *60*(4), 525–533. doi:10.1136/gut.2010.224436 [pii].

Kang, N. (2020). Mechanotransduction in liver diseases. *Seminars in Liver Disease*, *40*(1), 84–90. https://doi.org/10.1055/s-0039-3399502.

Kang, Y. B., Rawat, S., Cirillo, J., Bouchard, M., & Noh, H. M. (2013). Layered long-term co-culture of hepatocytes and endothelial cells on a transwell membrane: Toward engineering the liver sinusoid. *Biofabrication*, *5*(4). https://doi.org/10.1088/1758-5082/5/4/045008, 045008.

Kanninen, L. K., Harjumäki, R., Peltoniemi, P., Bogacheva, M. S., Salmi, T., Porola, P., et al. (2016). Laminin-511 and laminin-521-based matrices for efficient hepatic specification of human pluripotent stem cells. *Biomaterials*, *103*, 86–100. https://doi.org/10.1016/j.biomaterials.2016.06.054.

Kao, T., Labonne, T., Niclis, J. C., Chaurasia, R., Lokmic, Z., Qian, E., et al. (2016). GAPTrap: A simple expression system for pluripotent stem cells and their derivatives. *Stem Cell Reports*, *7*(3), 518–526. https://doi.org/10.1016/j.stemcr.2016.07.015.

Katsuda, T., Kawamata, M., Hagiwara, K., Takahashi, R. U., Yamamoto, Y., Camargo, F. D., et al. (2017). Conversion of terminally committed hepatocytes to culturable bipotent progenitor cells with regenerative capacity. *Cell Stem Cell*, *20*(1), 41–55. https://doi.org/10.1016/j.stem.2016.10.007.

Kawahara, T., Toso, C., Douglas, D. N., Nourbakhsh, M., Lewis, J. T., Tyrrell, D. L., et al. (2010). Factors affecting hepatocyte isolation, engraftment, and replication in an in vivo model. *Liver Transplantation*, *16*(8), 974–982. https://doi.org/10.1002/lt.22099.

Kim, J., Koo, B.-K., & Knoblich, J. A. (2020). Human organoids: Model systems for human biology and medicine. *Nature Reviews Molecular Cell Biology*, *21*(10), 571–584. https://doi.org/10.1038/s41580-020-0259-3.

Klaas, M., Kangur, T., Viil, J., Mäemets-Allas, K., Minajeva, A., Vadi, K., et al. (2016). The alterations in the extracellular matrix composition guide the repair of damaged liver tissue. *Scientific Reports*, *6*(1), 27398. https://doi.org/10.1038/srep27398.

Kleiman, R. J., & Engle, S. J. (2021). Human inducible pluripotent stem cells: Realization of initial promise in drug discovery. *Cell Stem Cell*, *28*(9), 1507–1515. https://doi.org/10.1016/j.stem.2021.08.002.

Koike, H., Iwasawa, K., Ouchi, R., Maezawa, M., Giesbrecht, K., Saiki, N., et al. (2019). Modelling human hepato-biliary-pancreatic organogenesis from the foregut–midgut boundary. *Nature*, *574*(7776), 112–116. https://doi.org/10.1038/s41586-019-1598-0.

Koui, Y., Kido, T., Ito, T., Oyama, H., Chen, S. W., Katou, Y., et al. (2017). An in vitro human liver model by iPSC-derived parenchymal and non-parenchymal cells. *Stem Cell Reports*, *9*(2), 490–498. https://doi.org/10.1016/j.stemcr.2017.06.010.

Kukla, D. A., Stoppel, W. L., Kaplan, D. L., & Khetani, S. R. (2020). Assessing the compatibility of primary human hepatocyte culture within porous silk sponges. *RSC Advances*, *10*(62), 37662–37674. https://doi.org/10.1039/D0RA04954A.

Lasli, S., Kim, H. J., Lee, K., Suurmond, C. E., Goudie, M., Bandaru, P., et al. (2019). A human liver-on-a-chip platform for modeling nonalcoholic fatty liver disease. *Adv Biosyst*, *3*(8). https://doi.org/10.1002/adbi.201900104, e1900104.

Lau, T. T., Ho, L. W., & Wang, D. A. (2013). Hepatogenesis of murine induced pluripotent stem cells in 3D micro-cavitary hydrogel system for liver regeneration. *Biomaterials*, *34*(28), 6659–6669. https://doi.org/10.1016/j.biomaterials.2013.05.034.

LeCluyse, E. L., Audus, K. L., & Hochman, J. H. (1994). Formation of extensive canalicular networks by rat hepatocytes cultured in collagen-sandwich configuration. *The American Journal of Physiology*, *266*(6 Pt 1), C1764–C1774. https://doi.org/10.1152/ajpcell.1994.266.6.C1764.

Lee, W., Cho, N.-J., Xiong, A., Glenn, J. S., & Frank, C. W. (2010). Hydrophobic nanoparticles improve permeability of cell-encapsulating poly(ethylene glycol) hydrogels while maintaining patternability. *Proceedings of the National Academy of Sciences*, *107*(48), 20709. https://doi.org/10.1073/pnas.1005211107.

Lee, H., Han, W., Kim, H., Ha, D.-H., Jang, J., Kim, B. S., et al. (2017). Development of liver decellularized extracellular matrix bioink for three-dimensional cell printing-based liver tissue engineering. *Biomacromolecules*, *18*(4), 1229–1237. https://doi.org/10.1021/acs.biomac.6b01908.

Lewis, P. L., Su, J., Yan, M., Meng, F., Glaser, S. S., Alpini, G. D., et al. (2018). Complex bile duct network formation within liver decellularized extracellular matrix hydrogels. *Scientific Reports*, *8*(1), 12220. https://doi.org/10.1038/s41598-018-30433-6.

Li, F., Cao, L., Parikh, S., & Zuo, R. (2020). Three-dimensional spheroids with primary human liver cells and differential roles of Kupffer cells in drug-induced liver injury. *Journal of Pharmaceutical Sciences*, *109*(6), 1912–1923. https://doi.org/10.1016/j.xphs.2020.02.021.

Li, X., George, S. M., Vernetti, L., Gough, A. H., & Taylor, D. L. (2018). A glass-based, continuously zonated and vascularized human liver acinus microphysiological system (vLAMPS) designed for experimental modeling of diseases and ADME/TOX. *Lab on a Chip*, *18*(17), 2614–2631. https://doi.org/10.1039/c8lc00418h.

Li, C. Y., Stevens, K. R., Schwartz, R. E., Alejandro, B. S., Huang, J. H., & Bhatia, S. N. (2014). Micropatterned cell-cell interactions enable functional encapsulation of primary hepatocytes in hydrogel microtissues. *Tissue Engineering. Part A*, *20*(15-16), 2200–2212. https://doi.org/10.1089/ten.TEA.2013.0667.

Liaw, C. Y., Ji, S., & Guvendiren, M. (2018). Engineering 3D hydrogels for personalized in vitro human tissue models. *Advanced Healthcare Materials, 7*(4). https://doi.org/10.1002/adhm.201701165.

Linti, C., Zipfel, A., Schenk, M., Dauner, M., Doser, M., Viebahn, R., et al. (2002). Cultivation of porcine hepatocytes in polyurethane nonwovens as part of a biohybrid liver support system. *The International Journal of Artificial Organs, 25*(10), 994–1000. https://doi.org/10.1177/039139880202501014.

Liu, X., LeCluyse, E. L., Brouwer, K. R., Gan, L. S., Lemasters, J. J., Stieger, B., et al. (1999). Biliary excretion in primary rat hepatocytes cultured in a collagen-sandwich configuration. *The American Journal of Physiology, 277*(1 Pt 1), G12–G21. https://doi.org/10.1152/ajpgi.1999.277.1.G12.

Liu, J., Li, P., Wang, L., Li, M., Ge, Z., Noordam, L., et al. (2021). Cancer-associated fibroblasts provide a stromal niche for liver cancer organoids that confers trophic effects and therapy resistance. *Cellular and Molecular Gastroenterology and Hepatology, 11*(2), 407–431. https://doi.org/10.1016/j.jcmgh.2020.09.003.

Llovet, J. M., Kelley, R. K., Villanueva, A., Singal, A. G., Pikarsky, E., Roayaie, S., et al. (2021). Hepatocellular carcinoma. *Nature Reviews. Disease Primers, 7*(1), 6. https://doi.org/10.1038/s41572-020-00240-3.

Lo, Y.-H., Karlsson, K., & Kuo, C. J. (2020). Applications of organoids for cancer biology and precision medicine. *Nature Cancer, 1*(8), 761–773. https://doi.org/10.1038/s43018-020-0102-y.

Lorenzini, S., Bird, T. G., Boulter, L., Bellamy, C., Samuel, K., Aucott, R., et al. (2010). Characterisation of a stereotypical cellular and extracellular adult liver progenitor cell niche in rodents and diseased human liver. *Gut, 59*(5), 645–654. 59/5/645 [pii].

Lu, H. F., Chua, K. N., Zhang, P. C., Lim, W. S., Ramakrishna, S., Leong, K. W., et al. (2005). Three-dimensional co-culture of rat hepatocyte spheroids and NIH/3T3 fibroblasts enhances hepatocyte functional maintenance. *Acta Biomaterialia, 1*(4), 399–410. https://doi.org/10.1016/j.actbio.2005.04.003.

Lu, J., Zhang, X., Li, J., Yu, L., Chen, E., Zhu, D., et al. (2016). A new fluidized bed bioreactor based on diversion-type microcapsule suspension for bioartificial liver systems. *PLoS One, 11*(2). https://doi.org/10.1371/journal.pone.0147376, e0147376.

Lynch, R. W., Hawley, C. A., Pellicoro, A., Bain, C. C., Iredale, J. P., & Jenkins, S. J. (2018). An efficient method to isolate Kupffer cells eliminating endothelial cell contamination and selective bias. *Journal of Leukocyte Biology, 104*(3), 579–586. https://doi.org/10.1002/JLB.1TA0517-169R.

Ma, L., Wu, Y., Li, Y., Aazmi, A., Zhou, H., Zhang, B., et al. (2020). Current advances on 3D-bioprinted liver tissue models. *Advanced Healthcare Materials, 9*(24). https://doi.org/10.1002/adhm.202001517, e2001517.

MacParland, S. A., Liu, J. C., Ma, X. Z., Innes, B. T., Bartczak, A. M., Gage, B. K., et al. (2018). Single cell RNA sequencing of human liver reveals distinct intrahepatic macrophage populations. *Nature Communications, 9*(1), 4383. https://doi.org/10.1038/s41467-018-06318-7.

Mak, K. M., & Mei, R. (2017). Basement membrane type IV collagen and laminin: An overview of their biology and value as fibrosis biomarkers of liver disease. *The Anatomical Record, 300*(8), 1371–1390. https://doi.org/10.1002/ar.23567.

Malinen, M. M., Kanninen, L. K., Corlu, A., Isoniemi, H. M., Lou, Y. R., Yliperttula, M. L., et al. (2014). Differentiation of liver progenitor cell line to functional organotypic cultures in 3D nanofibrillar cellulose and hyaluronan-gelatin hydrogels. *Biomaterials, 35*(19), 5110–5121. https://doi.org/10.1016/j.biomaterials.2014.03.020.

Mannaerts, I., Eysackers, N., Anne van Os, E., Verhulst, S., Roosens, T., Smout, A., et al. (2020). The fibrotic response of primary liver spheroids recapitulates in vivo hepatic stellate cell activation. *Biomaterials, 261*. https://doi.org/10.1016/j.biomaterials.2020.120335, 120335.

Manoilov, K. Y., Verkhusha, V. V., & Shcherbakova, D. M. (2021). A guide to the optogenetic regulation of endogenous molecules. *Nature Methods*, *18*(9), 1027–1037. https://doi.org/10.1038/s41592-021-01240-1.

Manzoni, G., Marinach, C., Topcu, S., Briquet, S., Grand, M., Tolle, M., et al. (2017). Plasmodium P36 determines host cell receptor usage during sporozoite invasion. *eLife*, *6*. https://doi.org/10.7554/eLife.25903.

March, S., Ramanan, V., Trehan, K., Ng, S., Galstian, A., Gural, N., et al. (2015). Micropatterned coculture of primary human hepatocytes and supportive cells for the study of hepatotropic pathogens. *Nature Protocols*, *10*(12), 2027–2053. https://doi.org/10.1038/nprot.2015.128.

Marongiu, F., Gramignoli, R., Dorko, K., Miki, T., Ranade, A. R., Paola Serra, M., et al. (2011). Hepatic differentiation of amniotic epithelial cells. *Hepatology*, *53*(5), 1719–1729. https://doi.org/10.1002/hep.24255.

Marsee, A., Roos, F. J. M., Verstegen, M. M. A., HPB Organoid Consortium, Gehart, H, de Koning, E, et al. (2021). Building consensus on definition and nomenclature of hepatic, pancreatic, and biliary organoids. *Cell Stem Cell*, *28*(5), 816–832. https://doi.org/10.1016/j.stem.2021.04.005.

Marshall, A., Rushbrook, S., Davies, S. E., Morris, L. S., Scott, I. S., Vowler, S. L., et al. (2005). Relation between hepatocyte G1 arrest, impaired hepatic regeneration, and fibrosis in chronic hepatitis C virus infection. *Gastroenterology*, *128*(1), 33–42. https://doi.org/10.1053/j.gastro.2004.09.076.

Martinez-Hernandez, A., & Amenta, P. S. (1993). The hepatic extracellular matrix. I. Components and distribution in normal liver. *Virchows Archiv. A, Pathological Anatomy and Histopathology*, *423*(1), 1–11. https://doi.org/10.1007/BF01606425.

Maschmeyer, I., Lorenz, A. K., Schimek, K., Hasenberg, T., Ramme, A. P., Hubner, J., et al. (2015). A four-organ-chip for interconnected long-term co-culture of human intestine, liver, skin and kidney equivalents. *Lab on a Chip*, *15*(12), 2688–2699. https://doi.org/10.1039/c5lc00392j.

Mattei, G., Magliaro, C., Pirone, A., & Ahluwalia, A. (2017). Decellularized human liver is too heterogeneous for designing a generic extracellular matrix mimic hepatic scaffold. *Artificial Organs*, *41*(12), E347–E355. https://doi.org/10.1111/aor.12925.

Mazza, G., Al-Akkad, W., Telese, A., Longato, L., Urbani, L., Robinson, B., et al. (2017). Rapid production of human liver scaffolds for functional tissue engineering by high shear stress oscillation-decellularization. *Scientific Reports*, *7*(1), 5534. https://doi.org/10.1038/s41598-017-05134-1.

Mazza, G., Rombouts, K., Hall, A. R., Urbani, L., Luong, T. V., Al-Akkad, W., et al. (2015). Decellularized human liver as a natural 3D-scaffold for liver bioengineering and transplantation. *Scientific Reports*, *7*(5), 13079. https://doi.org/10.1038/srep13079.

McCarron, S., Bathon, B., Conlon, D. M., Abbey, D., Rader, D. J., Gawronski, K., et al. (2021). Functional characterization of organoids derived from irreversibly damaged liver of patients with NASH. *Hepatology*. https://doi.org/10.1002/hep.31857 (Epublication ahead of print, 26 April 2021).

McClelland, R., Wauthier, E., Uronis, J., & Reid, L. (2008). Gradients in the liver's extracellular matrix chemistry from periportal to pericentral zones: Influence on human hepatic progenitors. *Tissue Engineering. Part A*, *14*(1), 59–70. https://doi.org/10.1089/ten.a.2007.0058.

McClelland, R., Wauthier, E., Zhang, L., Melhem, A., Schmelzer, E., Barbier, C., et al. (2008). Ex vivo conditions for self-replication of human hepatic stem cells. *Tissue Engineering Part C: Methods*, *14*(4), 341–351. https://doi.org/10.1089/ten.tec.2008.0073.

McCrary, M. W., Bousalis, D., Mobini, S., Song, Y. H., & Schmidt, C. E. (2020). Decellularized tissues as platforms for in vitro modeling of healthy and diseased tissues. Acta Biomaterialia, 111, 1–19. doi:https://doi.org/10.1016/j.actbio.2020.05.031.

Michalopoulos, G., Russell, F., & Biles, C. (1979). Primary cultures of hepatocytes on human fibroblasts. *In Vitro, 15*(10), 796–806. https://doi.org/10.1007/bf02618306.

Mirmalek-Sani, S. H., Sullivan, D. C., Zimmerman, C., Shupe, T. D., & Petersen, B. E. (2013). Immunogenicity of decellularized porcine liver for bioengineered hepatic tissue. *The American Journal of Pathology, 183*(2), 558–565. https://doi.org/10.1016/j.ajpath.2013.05.002.

Miyagiwa, M., Ichida, T., Tokiwa, T., Sato, J., & Sasaki, H. (1989). A new human cholangiocellular carcinoma cell line (HuCC-T1) producing carbohydrate antigen 19/9 in serum-free medium. *In Vitro Cellular & Developmental Biology, 25*(6), 503–510. https://doi.org/10.1007/BF02623562.

Mizuno, K., & Ueno, Y. (2017). Autonomic nervous system and the liver. *Hepatology Research, 47*(2), 160–165. https://doi.org/10.1111/hepr.12760.

Moghe, P. V., Berthiaume, F., Ezzell, R. M., Toner, M., Tompkins, R. G., & Yarmush, M. L. (1996). Culture matrix configuration and composition in the maintenance of hepatocyte polarity and function. *Biomaterials, 17*(3), 373–385. https://doi.org/10.1016/0142-9612(96)85576-1.

Molina, D. K., & DiMaio, V. J. (2012). Normal organ weights in men: Part II—The brain, lungs, liver, spleen, and kidneys. *The American Journal of Forensic Medicine and Pathology, 33*(4), 368–372. https://doi.org/10.1097/PAF.0b013e31823d29ad.

Morin, O., & Normand, C. (1986). Long-term maintenance of hepatocyte functional activity in co-culture: Requirements for sinusoidal endothelial cells and dexamethasone. *Journal of Cellular Physiology, 129*(1), 103–110. https://doi.org/10.1002/jcp.1041290115.

Ng, S. S., Saeb-Parsy, K., Blackford, S. J. I., Segal, J. M., Serra, M. P., Horcas-Lopez, M., et al. (2018). Human iPS derived progenitors bioengineered into liver organoids using an inverted colloidal crystal poly (ethylene glycol) scaffold. *Biomaterials, 182,* 299–311. https://doi.org/10.1016/j.biomaterials.2018.07.043.

Nishii, K., Brodin, E., Renshaw, T., Weesner, R., Moran, E., Soker, S., et al. (2018). Shear stress upregulates regeneration-related immediate early genes in liver progenitors in 3D ECM-like microenvironments. *Journal of Cellular Physiology, 233*(5), 4272–4281. https://doi.org/10.1002/jcp.26246.

No, D. Y., Lee, S. A., Choi, Y. Y., Park, D., Jang, J. Y., Kim, D. S., et al. (2012). Functional 3D human primary hepatocyte spheroids made by co-culturing hepatocytes from partial hepatectomy specimens and human adipose-derived stem cells. *PLoS One, 7*(12). https://doi.org/10.1371/journal.pone.0050723, e50723.

Nugraha, B., Hong, X., Mo, X., Tan, L., Zhang, W., Chan, P.-M., et al. (2011). Galactosylated cellulosic sponge for multi-well drug safety testing. *Biomaterials, 32,* 6982–6994. https://doi.org/10.1016/j.biomaterials.2011.05.087.

Nwosu, Z. C., Battello, N., Rothley, M., Piorońska, W., Sitek, B., Ebert, M. P., et al. (2018). Liver cancer cell lines distinctly mimic the metabolic gene expression pattern of the corresponding human tumours. *Journal of Experimental & Clinical Cancer Research, 37*(1), 211. https://doi.org/10.1186/s13046-018-0872-6.

Nykonenko, A., Vávra, P., & Zonča, P. (2017). Anatomic peculiarities of pig and human liver. *Experimental and Clinical Transplantation, 15*(1), 21–26. https://doi.org/10.6002/ect.2016.0099.

Ober, E. A., & Lemaigre, F. P. (2018). Development of the liver: Insights into organ and tissue morphogenesis. *Journal of Hepatology, 68*(5), 1049–1062. https://doi.org/10.1016/j.jhep.2018.01.005.

O'Brien, F. J. (2011). Biomaterials & scaffolds for tissue engineering. *Materials Today, 14*(3), 88–95. https://doi.org/10.1016/S1369-7021(11)70058-X.

Ogata, I., Mochida, S., Tomiya, T., & Fujiwara, K. (1991). Minor contribution of hepatocytes to collagen production in normal and early fibrotic rat livers. *Hepatology, 14*(2), 361–367.

Olsen, A. L., Bloomer, S. A., Chan, E. P., Gaça, M. D. A., Georges, P. C., Sackey, B., et al. (2011). Hepatic stellate cells require a stiff environment for myofibroblastic differentiation. *American Journal of Physiology. Gastrointestinal and Liver Physiology, 301*(1), G110–G118. https://doi.org/10.1152/ajpgi.00412.2010.

Ouchi, R., Togo, S., Kimura, M., Shinozawa, T., Koido, M., Koike, H., et al. (2019). Modeling steatohepatitis in humans with pluripotent stem cell-derived organoids. *Cell Metabolism, 30*(2), 374–384 e376. https://doi.org/10.1016/j.cmet.2019.05.007.

Parent, R., Durantel, D., Lahlali, T., Salle, A., Plissonnier, M. L., DaCosta, D., et al. (2014). An immortalized human liver endothelial sinusoidal cell line for the study of the pathobiology of the liver endothelium. *Biochemical and Biophysical Research Communications, 450*(1), 7–12. https://doi.org/10.1016/j.bbrc.2014.05.038.

Parent, R., Marion, M. J., Furio, L., Trépo, C., & Petit, M. A. (2004). Origin and characterization of a human bipotent liver progenitor cell line. *Gastroenterology, 126*(4), 1147–1156. https://doi.org/10.1053/j.gastro.2004.01.002.

Park, K.-M., Hussein, K. H., Hong, S.-H., Ahn, C., Yang, S.-R., Park, S.-M., et al. (2016). Decellularized liver extracellular matrix as promising tools for transplantable bioengineered liver promotes hepatic lineage commitments of induced pluripotent stem cells. *Tissue Engineering Part A, 22*(5-6), 449–460. https://doi.org/10.1089/ten.tea.2015.0313.

Payen, V. L., Lavergne, A., Alevra Sarika, N., Colonval, M., Karim, L., Deckers, M., et al. (2021). Single-cell RNA sequencing of human liver reveals hepatic stellate cell heterogeneity. *JHEP Reports, 3*(3). https://doi.org/10.1016/j.jhep.2021.100278, 100278.

Peng, W. C., Logan, C. Y., Fish, M., Anbarchian, T., Aguisanda, F., Alvarez-Varela, A., et al. (2018). Inflammatory cytokine TNFalpha promotes the long-term expansion of primary hepatocytes in 3D culture. *Cell, 175*(6), 1607–1619. e1615 https://doi.org/10.1016/j.cell.2018.11.012.

Perez, R. A., Jung, C.-R., & Kim, H.-W. (2017). Biomaterials and culture technologies for regenerative therapy of liver tissue. *Advanced Healthcare Materials, 6*(2), 1600791. https://doi.org/10.1002/adhm.201600791.

Planas-Paz, L., Orsini, V., Boulter, L., Calabrese, D., Pikiolek, M., Nigsch, F., et al. (2016). The RSPO-LGR4/5-ZNRF3/RNF43 module controls liver zonation and size. *Nature Cell Biology, 18*(5), 467–479. https://doi.org/10.1038/ncb3337.

Poisson, J., Lemoinne, S., Boulanger, C., Durand, F., Moreau, R., Valla, D., et al. (2017). Liver sinusoidal endothelial cells: Physiology and role in liver diseases. *Journal of Hepatology, 66*(1), 212–227. https://doi.org/10.1016/j.jhep.2016.07.009.

Powell, E. E., Wong, V. W., & Rinella, M. (2021). Non-alcoholic fatty liver disease. *Lancet, 397*(10290), 2212–2224. https://doi.org/10.1016/s0140-6736(20)32511-3.

Rajendran, D., Hussain, A., Yip, D., Parekh, A., Shrirao, A., & Cho, C. H. (2017). Long-term liver-specific functions of hepatocytes in electrospun chitosan nanofiber scaffolds coated with fibronectin. *Journal of Biomedical Materials Research. Part A, 105*(8), 2119–2128. https://doi.org/10.1002/jbm.a.36072.

Ramachandran, P., Dobie, R., Wilson-Kanamori, J. R., Dora, E. F., Henderson, B. E. P., Luu, N. T., et al. (2019). Resolving the fibrotic niche of human liver cirrhosis at single-cell level. *Nature, 575*(7783), 512–518. https://doi.org/10.1038/s41586-019-1631-3.

Ramachandran, S. D., Schirmer, K., Munst, B., Heinz, S., Ghafoory, S., Wolfl, S., et al. (2015). In vitro generation of functional liver organoid-like structures using adult human cells. *PLoS One, 10*(10). https://doi.org/10.1371/journal.pone.0139345, e0139345.

Ramli, M. N. B., Lim, Y. S., Koe, C. T., Demircioglu, D., Tng, W., Gonzales, K. A. U., et al. (2020). Human pluripotent stem cell-derived organoids as models of liver disease. *Gastroenterology, 159*(4), 1471–1486.e1412. https://doi.org/10.1053/j.gastro.2020.06.010.

Raven, A., & Forbes, S. J. (2018). Hepatic progenitors in liver regeneration. *Journal of Hepatology, 69*(6), 1394–1395. https://doi.org/10.1016/j.jhep.2018.03.004.

Raven, A., Lu, W. Y., Man, T. Y., Ferreira-Gonzalez, S., O'Duibhir, E., Dwyer, B. J., et al. (2017). Cholangiocytes act as facultative liver stem cells during impaired hepatocyte regeneration. *Nature, 547*(7663), 350–354. https://doi.org/10.1038/nature23015.

Ren, H., Shi, X., Tao, L., Xiao, J., Han, B., Zhang, Y., et al. (2013). Evaluation of two decellularization methods in the development of a whole-organ decellularized rat liver scaffold. *Liver International, 33*(3), 448–458. https://doi.org/10.1111/liv.12088.

Rennert, K., Steinborn, S., Groger, M., Ungerbock, B., Jank, A. M., Ehgartner, J., et al. (2015). A microfluidically perfused three dimensional human liver model. *Biomaterials, 71*, 119–131. https://doi.org/10.1016/j.biomaterials.2015.08.043.

Rimland, C. A., Tilson, S. G., Morell, C. M., Tomaz, R. A., Lu, W.-Y., Adams, S. E., et al. (2021). Regional differences in human biliary tissues and corresponding in vitro–derived organoids. *Hepatology, 73*(1), 247–267. https://doi.org/10.1002/hep.31252.

Rose, K. A., Holman, N. S., Green, A. M., Andersen, M. E., & LeCluyse, E. L. (2016). Co-culture of hepatocytes and Kupffer cells as an in vitro model of inflammation and drug-induced hepatotoxicity. *Journal of Pharmaceutical Sciences, 105*(2), 950–964. https://doi.org/10.1016/S0022-3549(15)00192-6.

Roskams, T., Rosenbaum, J., De Vos, R., David, G., & Desmet, V. (1996). Heparan sulfate proteoglycan expression in chronic cholestatic human liver diseases. *Hepatology, 24*(3), 524–532. https://doi.org/10.1053/jhep.1996.v24.pm0008781318.

Ruoß, M., Häussling, V., Schügner, F., Olde Damink, L. H. H., Lee, S. M. L., Ge, L., et al. (2018). A standardized collagen-based scaffold improves human hepatocyte shipment and allows metabolic studies over 10 days. *Bioengineering (Basel), 5*(4), 86. https://doi.org/10.3390/bioengineering5040086.

Saheli, M., Sepantafar, M., Pournasr, B., Farzaneh, Z., Vosough, M., Piryaei, A., et al. (2018). Three-dimensional liver-derived extracellular matrix hydrogel promotes liver organoids function. *Journal of Cellular Biochemistry, 119*(6), 4320–4333. https://doi.org/10.1002/jcb.26622.

Saito, Y., Muramatsu, T., Kanai, Y., Ojima, H., Sukeda, A., Hiraoka, N., et al. (2019). Establishment of patient-derived organoids and drug screening for biliary tract carcinoma. *Cell Reports, 27*(4), 1265–1276 e1264. https://doi.org/10.1016/j.celrep.2019.03.088.

Sampaziotis, F., Cardoso de Brito, M., Madrigal, P., Bertero, A., Saeb-Parsy, K., Soares, F. A. C., et al. (2015). Cholangiocytes derived from human induced pluripotent stem cells for disease modeling and drug validation. *Nature Biotechnology, 33*(8), 845–852. https://doi.org/10.1038/nbt.3275.

Sampaziotis, F., Justin, A. W., Tysoe, O. C., Sawiak, S., Godfrey, E. M., Upponi, S. S., et al. (2017). Reconstruction of the mouse extrahepatic biliary tree using primary human extrahepatic cholangiocyte organoids. *Nature Medicine, 23*(8), 954–963. https://doi.org/10.1038/nm.4360.

Sampaziotis, F., Muraro, D., Tysoe, O. C., Sawiak, S., Beach, T. E., Godfrey, E. M., et al. (2021). Cholangiocyte organoids can repair bile ducts after transplantation in the human liver. *Science, 371*(6531), 839–846. https://doi.org/10.1126/science.aaz6964.

Schepers, A., Li, C., Chhabra, A., Seney, B. T., & Bhatia, S. (2016). Engineering a perfusable 3D human liver platform from iPS cells. *Lab on a Chip, 16*(14), 2644–2653. https://doi.org/10.1039/c6lc00598e.

Schneeberger, K., Sanchez-Romero, N., Ye, S., van Steenbeek, F. G., Oosterhoff, L. A., Pla Palacin, I., et al. (2020). Large-scale production of LGR5-positive bipotential human liver stem cells. *Hepatology, 72*(1), 257–270. https://doi.org/10.1002/hep.31037.

Sefried, S., Haring, H., Weigert, C., & Eckstein, S. (2018). Suitability of hepatocyte cell lines HepG2, AML12 and THLE-2 for investigation of insulin signalling and hepatokine gene expression. *Open Biology, 8*. https://doi.org/10.1098/rsob.180147, 180147.

Sellaro, T. L., Ranade, A., Faulk, D. M., McCabe, G. P., Dorko, K., Badylak, S. F., et al. (2010). Maintenance of human hepatocyte function in vitro by liver-derived extracellular matrix gels. *Tissue Engineering. Part A, 16*(3), 1075–1082. https://doi.org/10.1089/ten.TEA.2008.0587.

Shaheen, M. F., Joo, D. J., Ross, J. J., Anderson, B. D., Chen, H. S., Huebert, R. C., et al. (2020). Sustained perfusion of revascularized bioengineered livers heterotopically transplanted into immunosuppressed pigs. *Nature Biomedical Engineering, 4*(4), 437–445. https://doi.org/10.1038/s41551-019-0460-x.

Shupe, T., Williams, M., Brown, A., Willenberg, B., & Petersen, B. E. (2010). Method for the decellularization of intact rat liver. *Organogenesis, 6*(2), 134–136. https://doi.org/10.4161/org.6.2.11546.

Sibille, J.-C., Kondo, H., & Aisen, P. (1988). Interactions between isolated hepatocytes and Kupffer cells in iron metabolism: A possible role for ferritin as an iron carrier protein. *Hepatology, 8*(2), 296–301. https://doi.org/10.1002/hep.1840080218.

Siltanen, C., Diakatou, M., Lowen, J., Haque, A., Rahimian, A., Stybayeva, G., et al. (2017). One step fabrication of hydrogel microcapsules with hollow core for assembly and cultivation of hepatocyte spheroids. *Acta Biomaterialia, 50*, 428–436. https://doi.org/10.1016/j.actbio.2017.01.010.

Sinturel, F., Gerber, A., Mauvoisin, D., Wang, J., Gatfield, D., Stubblefield, J. J., et al. (2017). Diurnal oscillations in liver mass and cell size accompany ribosome assembly cycles. *Cell, 169*(4), 651–663 e614. https://doi.org/10.1016/j.cell.2017.04.015.

Sison-Young, R. L. C., Mitsa, D., Jenkins, R. E., Mottram, D., Alexandre, E., Richert, L., et al. (2015). Comparative proteomic characterization of 4 human liver-derived single cell culture models reveals significant variation in the capacity for drug disposition, bioactivation, and detoxication. *Toxicological Sciences, 147*(2), 412–424. https://doi.org/10.1093/toxsci/kfv136.

Si-Tayeb, K., Lemaigre, F. P., & Duncan, S. A. (2010). Organogenesis and development of the liver. *Developmental Cell, 18*(2), 175–189. https://doi.org/10.1016/j.devcel.2010.01.011.

Slaughter, V. L., Rumsey, J. W., Boone, R., Malik, D., Cai, Y., Sriram, N. N., et al. (2021). Validation of an adipose-liver human-on-a-chip model of NAFLD for preclinical therapeutic efficacy evaluation. *Scientific Reports, 11*(1), 13159. https://doi.org/10.1038/s41598-021-92264-2.

Soltantabar, P., Calubaquib, E. L., Mostafavi, E., Ghazavi, A., & Stefan, M. C. (2021). Heart/liver-on-a-chip as a model for the evaluation of cardiotoxicity induced by chemotherapies. *Organs-on-a-Chip, 3.* https://doi.org/10.1016/j.ooc.2021.100008, 100008.

Sorrentino, G., Rezakhani, S., Yildiz, E., Nuciforo, S., Heim, M. H., Lutolf, M. P., et al. (2020). Mechano-modulatory synthetic niches for liver organoid derivation. *Nature Communications, 11*(1), 3416. https://doi.org/10.1038/s41467-020-17161-0.

Soto-Gutiérrez, A., Navarro-Álvarez, N., Zhao, D., Rivas-Carrillo, J. D., Lebkowski, J., Tanaka, N., et al. (2007). Differentiation of mouse embryonic stem cells to hepatocyte-like cells by co-culture with human liver nonparenchymal cell lines. *Nature Protocols, 2*(2), 347–356. https://doi.org/10.1038/nprot.2007.18.

Soto-Gutierrez, A., Zhang, L., Medberry, C., Fukumitsu, K., Faulk, D., Jiang, H., et al. (2011). A whole-organ regenerative medicine approach for liver replacement. *Tissue Engineering. Part C, Methods, 17*(6), 677–686. https://doi.org/10.1089/ten.tec.2010.0698.

Stevens, K. R., Scull, M. A., Ramanan, V., Fortin, C. L., Chaturvedi, R. R., Knouse, K. A., et al. (2017). In situ expansion of engineered human liver tissue in a mouse model of chronic liver disease. *Science Translational Medicine, 9*(399). https://doi.org/10.1126/scitranslmed.aah5505.

Strauss, O., Phillips, A., Ruggiero, K., Bartlett, A., & Dunbar, P. R. (2017). Immunofluorescence identifies distinct subsets of endothelial cells in the human liver. *Scientific Reports*, 7, 44356. https://doi.org/10.1038/srep44356.

Struecker, B., Hillebrandt, K. H., Voitl, R., Butter, A., Schmuck, R. B., Reutzel-Selke, A., et al. (2015). Porcine liver decellularization under oscillating pressure conditions: A technical refinement to improve the homogeneity of the decellularization process. *Tissue Engineering. Part C, Methods*, 21(3), 303–313. https://doi.org/10.1089/ten. TEC.2014.0321.

Su, T., Yang, Y., Lai, S., Jeong, J., Jung, Y., McConnell, M., et al. (2021). Single-cell transcriptomics reveals zone-specific alterations of liver sinusoidal endothelial cells in cirrhosis. *Cellular and Molecular Gastroenterology and Hepatology*, 11(4), 1139–1161. https://doi. org/10.1016/j.jcmgh.2020.12.007.

Sue, C. H., Hal, H. H., Jim, B., Taylor, L., & Darlington, G. J. (1992). SK HEP-1: A human cell line of endothelial origin. *In Vitro Cellular & Developmental Biology*, 28A(2), 136–142. https://doi.org/10.1007/BF02631017.

Tabibian, J. H., Trussoni, C. E., O'Hara, S. P., Splinter, P. L., Heimbach, J. K., & LaRusso, N. F. (2014). Characterization of cultured cholangiocytes isolated from livers of patients with primary sclerosing cholangitis. *Laboratory Investigation*, 94(10), 1126–1133. https://doi.org/10.1038/labinvest.2014.94.

Takahashi, K., Tanabe, K., Ohnuki, M., Narita, M., Ichisaka, T., Tomoda, K., et al. (2007). Induction of pluripotent stem cells from adult human fibroblasts by defined factors. *Cell*, 131(5), 861–872. https://doi.org/10.1016/j.cell.2007.11.019.

Takayama, K., Mitani, S., Nagamoto, Y., Sakurai, F., Tachibana, M., Taniguchi, Y., et al. (2016). Laminin 411 and 511 promote the cholangiocyte differentiation of human induced pluripotent stem cells. *Biochemical and Biophysical Research Communications*, 474(1), 91–96. https://doi.org/10.1016/j.bbrc.2016.04.075.

Takebe, T., Sekine, K., Enomura, M., Koike, H., Kimura, M., Ogaeri, T., et al. (2013). Vascularized and functional human liver from an iPSC-derived organ bud transplant. *Nature*, 499(7459), 481–484. https://doi.org/10.1038/nature12271.

Takeishi, K., Collin de l'Hortet, A., Wang, Y., Handa, K., Guzman-Lepe, J., Matsubara, K., et al. (2020). Assembly and function of a bioengineered human liver for transplantation generated solely from induced pluripotent stem cells. *Cell Reports*, 31(9). https://doi.org/ 10.1016/j.celrep.2020.107711, 107711.

Tanaka, M., & Iwakiri, Y. (2016). The hepatic lymphatic vascular system: Structure, function, markers, and lymphangiogenesis. *Cellular and Molecular Gastroenterology and Hepatology*, 2(6), 733–749. https://doi.org/10.1016/j.jcmgh.2016.09.002.

Tang, X.-P., Zhang, M., Yang, X., Chen, L.-M., & Zeng, Y. (2006). Differentiation of human umbilical cord blood stem cells into hepatocytes in vivo and in vitro. *World Journal of Gastroenterology*, 12(25), 4014–4019. https://doi.org/10.3748/wjg.v12.i25. 4014.

Tanimizu, N., Ichinohe, N., Sasaki, Y., Itoh, T., Sudo, R., Yamaguchi, T., et al. (2021). Generation of functional liver organoids on combining hepatocytes and cholangiocytes with hepatobiliary connections ex vivo. *Nature Communications*, 12(1), 3390. https://doi. org/10.1038/s41467-021-23575-1.

Tanimizu, N., Miyajima, A., & Mostov, K. E. (2007). Liver progenitor cells develop cholangiocyte-type epithelial polarity in three-dimensional culture. *Molecular Biology of the Cell*, 18(4), 1472–1479. 10.1091/mbc.E06-09-0848.

Tasnim, F., Xing, J., Huang, X., Mo, S., Wei, X., Tan, M. H., et al. (2019). Generation of mature kupffer cells from human induced pluripotent stem cells. *Biomaterials*, 192, 377–391. https://doi.org/10.1016/j.biomaterials.2018.11.016.

Tatrai, P., Dudas, J., Batmunkh, E., Mathe, M., Zalatnai, A., Schaff, Z., et al. (2006). Agrin, a novel basement membrane component in human and rat liver, accumulates in cirrhosis and hepatocellular carcinoma. *Laboratory Investigation*, *86*(11), 1149–1160. https://doi.org/10.1038/labinvest.3700475.

Teratani, T., Quinn, G., Yamamoto, Y., Sato, T., Yamanokuchi, H., Asari, A., et al. (2005). Long-term maintenance of liver-specific functions in cultured ES cell-derived hepatocytes with hyaluronan sponge. *Cell Transplantation*, *14*(9), 629–635. https://doi.org/10.3727/000000005783982611.

Terry, C., Dhawan, A., Mitry, R. R., Lehec, S. C., & Hughes, R. D. (2010). Optimization of the cryopreservation and thawing protocol for human hepatocytes for use in cell transplantation. *Liver Transplantation*, *16*(2), 229–237. https://doi.org/10.1002/lt.21983.

Thomson, J. A., Itskovitz-Eldor, J., Shapiro, S. S., Waknitz, M. A., Swiergiel, J. J., Marshall, V. S., et al. (1998). Embryonic stem cell lines derived from human blastocysts. *Science*, *282*(5391), 1145–1147. https://doi.org/10.1126/science.282.5391.1145.

Tonon, F., Giobbe, G. G., Zambon, A., Luni, C., Gagliano, O., Floreani, A., et al. (2019). In vitro metabolic zonation through oxygen gradient on a chip. *Scientific Reports*, *9*(1), 13557. https://doi.org/10.1038/s41598-019-49412-6.

Török, E., Lutgehetmann, M., Bierwolf, J., Melbeck, S., Düllmann, J., Nashan, B., et al. (2011). Primary human hepatocytes on biodegradable poly(l-lactic acid) matrices: A promising model for improving transplantation efficiency with tissue engineering. *Liver Transplantation*, *17*(2), 104–114. https://doi.org/10.1002/lt.22200.

Tostoes, R. M., Leite, S. B., Serra, M., Jensen, J., Bjorquist, P., Carrondo, M. J., et al. (2012). Human liver cell spheroids in extended perfusion bioreactor culture for repeated-dose drug testing. *Hepatology*, *55*(4), 1227–1236. https://doi.org/10.1002/hep.24760.

Trefts, E., Gannon, M., & Wasserman, D. H. (2017). The liver. *Current Biology*, *27*(21), R1147–R1151. https://doi.org/10.1016/j.cub.2017.09.019.

Tseng, S. C., Lee, P. C., Ells, P. F., Bissell, D. M., Smuckler, E. A., & Stern, R. (1982). Collagen production by rat hepatocytes and sinusoidal cells in primary monolayer culture. *Hepatology*, *2*(1), 13–18. https://doi.org/10.1002/hep.1840020104.

Tysoe, O. C., Justin, A. W., Brevini, T., Chen, S. E., Mahbubani, K. T., Frank, A. K., et al. (2019). Isolation and propagation of primary human cholangiocyte organoids for the generation of bioengineered biliary tissue. *Nature Protocols*, *14*(6), 1884–1925. https://doi.org/10.1038/s41596-019-0168-0.

Underhill, G. H., Chen, A. A., Albrecht, D. R., & Bhatia, S. N. (2007). Assessment of hepatocellular function within PEG hydrogels. *Biomaterials*, *28*(2), 256–270.

Underhill, G. H., & Khetani, S. R. (2018). Bioengineered liver models for drug testing and cell differentiation studies. *Cellular and Molecular Gastroenterology and Hepatology*, *5*(3), 426–439 e421. https://doi.org/10.1016/j.jcmgh.2017.11.012.

Uyama, N., Shimahara, Y., Kawada, N., Seki, S., Okuyama, H., Iimuro, Y., et al. (2002). Regulation of cultured rat hepatocyte proliferation by stellate cells. *Journal of Hepatology*, *36*(5), 590–599. https://doi.org/10.1016/S0168-8278(02)00023-5.

Uygun, B. E., Soto-Gutierrez, A., Yagi, H., Izamis, M. L., Guzzardi, M. A., Shulman, C., et al. (2010). Organ reengineering through development of a transplantable recellularized liver graft using decellularized liver matrix. *Nature Medicine*, *16*(7), 814–820. https://doi.org/10.1038/nm.2170.

Velasco, V., Shariati, S. A., & Esfandyarpour, R. (2020). Microtechnology-based methods for organoid models. *Microsystems & Nanoengineering*, *6*(1), 76. https://doi.org/10.1038/s41378-020-00185-3.

Verhulst, S., Roskams, T., Sancho-Bru, P., & van Grunsven, L. A. (2019). Meta-analysis of human and mouse biliary epithelial cell gene profiles. *Cell*, *8*(10), 1117. https://doi.org/10.3390/cells8101117.

Verstegen, M. M. A., Willemse, J., van den Hoek, S., Kremers, G.-J., Luider, T. M., van Huizen, N. A., et al. (2017). Decellularization of whole human liver grafts using controlled perfusion for transplantable organ bioscaffolds. *Stem Cells and Development, 26*(18), 1304–1315. https://doi.org/10.1089/scd.2017.0095.

Vunjak-Novakovic, G., Ronaldson-Bouchard, K., & Radisic, M. (2021). Organs-on-a-chip models for biological research. *Cell, 184*(18), 4597–4611. https://doi.org/10.1016/j.cell.2021.08.005.

Wahlicht, T., Vièyres, G., Bruns, S. A., Meumann, N., Büning, H., Hauser, H., et al. (2020). Controlled functional zonation of hepatocytes in vitro by engineering of Wnt signaling. *ACS Synthetic Biology, 9*(7), 1638–1649. https://doi.org/10.1021/acssynbio.9b00435.

Wang, Y., Cui, C. B., Yamauchi, M., Miguez, P., Roach, M., Malavarca, R., et al. (2011). Lineage restriction of human hepatic stem cells to mature fates is made efficient by tissue-specific biomatrix scaffolds. *Hepatology, 53*(1), 293–305. https://doi.org/10.1002/hep.24012.

Wang, S., Nagrath, D., Chen, P. C., Berthiaume, F., & Yarmush, M. L. (2008). Three-dimensional primary hepatocyte culture in synthetic self-assembling peptide hydrogel. *Tissue Engineering. Part A, 14*(2), 227–236. https://doi.org/10.1089/tea.2007.0143.

Wang, Y., Song, L., Liu, M., Ge, R., Zhou, Q., Liu, W., et al. (2018). A proteomics landscape of circadian clock in mouse liver. *Nature Communications, 9*(1), 1553. https://doi.org/10.1038/s41467-018-03898-2.

Wang, L., Wang, X., Xie, G., Wang, L., Hill, C. K., & DeLeve, L. D. (2012). Liver sinusoidal endothelial cell progenitor cells promote liver regeneration in rats. *The Journal of Clinical Investigation, 122*(4), 1567–1573. https://doi.org/10.1172/JCI58789.

Ware, B. R., Berger, D. R., & Khetani, S. R. (2015). Prediction of drug-induced liver injury in micropatterned co-cultures containing iPSC-derived human hepatocytes. *Toxicological Sciences, 145*(2), 252–262. https://doi.org/10.1093/toxsci/kfv048.

Ware, B. R., Durham, M. J., Monckton, C. P., & Khetani, S. R. (2018). A cell culture platform to maintain long-term phenotype of primary human hepatocytes and endothelial cells. *Cellular and Molecular Gastroenterology and Hepatology, 5*(3), 187–207. https://doi.org/10.1016/j.jcmgh.2017.11.007.

Ware, B. R., & Khetani, S. R. (2017). Engineered liver platforms for different phases of drug development. *Trends in Biotechnology, 35*(2), 172–183. https://doi.org/10.1016/j.tibtech.2016.08.001.

Watanabe, M., Zemack, H., Johansson, H., Hagbard, L., Jorns, C., Li, M., et al. (2016). Maintenance of hepatic functions in primary human hepatocytes cultured on xeno-free and chemical defined human recombinant laminins. *PLoS One, 11*(9). https://doi.org/10.1371/journal.pone.0161383, e0161383.

Weeks, C. A., Newman, K., Turner, P. A., Rodysill, B., Hickey, R. D., Nyberg, S. L., et al. (2013). Suspension culture of hepatocyte-derived reporter cells in presence of albumin to form stable three-dimensional spheroids. *Biotechnology and Bioengineering, 110*(9), 2548–2555. https://doi.org/10.1002/bit.24899.

Wei, Y., Wang, Y. G., Jia, Y., Li, L., Yoon, J., Zhang, S., et al. (2021). Liver homeostasis is maintained by midlobular zone 2 hepatocytes. *Science, 371*(6532). https://doi.org/10.1126/science.abb1625.

Wen, Y., Lambrecht, J., Ju, C., & Tacke, F. (2021). Hepatic macrophages in liver homeostasis and diseases-diversity, plasticity and therapeutic opportunities. *Cellular & Molecular Immunology, 18*(1), 45–56. https://doi.org/10.1038/s41423-020-00558-8.

Willemse, J., Verstegen, M. M. A., Vermeulen, A., Schurink, I. J., Roest, H. P., van der Laan, L. J. W., et al. (2020). Fast, robust and effective decellularization of whole human livers using mild detergents and pressure controlled perfusion. *Materials Science and Engineering: C, 108*. https://doi.org/10.1016/j.msec.2019.110200, 110200.

Wu, Y., Wenger, A., Golzar, H., & Tang, X. (2020). 3D bioprinting of bicellular liver lobule-mimetic structures via microextrusion of cellulose nanocrystal-incorporated shear-thinning bioink. *Scientific Reports, 10*(1), 20648. https://doi.org/10.1038/s41598-020-77146-3.

Wu, F., Wu, D., Ren, Y., Huang, Y., Feng, B., Zhao, N., et al. (2019). Generation of hepatobiliary organoids from human induced pluripotent stem cells. *Journal of Hepatology, 70*(6), 1145–1158. https://doi.org/10.1016/j.jhep.2018.12.028.

Xia, Y., Carpentier, A., Cheng, X., Block, P. D., Zhao, Y., Zhang, Z., et al. (2017). Human stem cell-derived hepatocytes as a model for hepatitis B virus infection, spreading and virus-host interactions. *Journal of Hepatology, 66*(3), 494–503. https://doi.org/10.1016/j.jhep.2016.10.009.

Xia, T., Zhao, R., Feng, F., & Yang, L. (2020). The effect of matrix stiffness on human hepatocyte migration and function—An in vitro research. *Polymers (Basel), 12*(9). https://doi.org/10.3390/polym12091903.

Xie, G., Choi, S. S., Syn, W. K., Michelotti, G. A., Swiderska, M., Karaca, G., et al. (2013). Hedgehog signalling regulates liver sinusoidal endothelial cell capillarisation. *Gut, 62*(2), 299–309. https://doi.org/10.1136/gutjnl-2011-301494.

Xu, L., Hui, A. Y., Albanis, E., Arthur, M. J., O'Byrne, S. M., Blaner, W. S., et al. (2005). Human hepatic stellate cell lines, LX-1 and LX-2: New tools for analysis of hepatic fibrosis. *Gut, 54*(1), 142–151. https://doi.org/10.1136/gut.2004.042127.

Yagi, H., Fukumitsu, K., Fukuda, K., Kitago, M., Shinoda, M., Obara, H., et al. (2013). Human-scale whole-organ bioengineering for liver transplantation: A regenerative medicine approach. *Cell Transplantation, 22*(2), 231–242. https://doi.org/10.3727/096368912x654939.

Yamashita, T., & Wang, X. W. (2013). Cancer stem cells in the development of liver cancer. *The Journal of Clinical Investigation, 123*(5), 1911–1918. https://doi.org/10.1172/JCI66024.

Yan, H., Zhong, G., Xu, G., He, W., Jing, Z., Gao, Z., et al. (2012). Sodium taurocholate cotransporting polypeptide is a functional receptor for human hepatitis B and D virus. *eLife, 1*, e00049. https://doi.org/10.7554/eLife.00049.

Yap, K. K., Dingle, A. M., Palmer, J. A., Dhillon, R. S., Lokmic, Z., Penington, A. J., et al. (2013). Enhanced liver progenitor cell survival and differentiation in vivo by spheroid implantation in a vascularized tissue engineering chamber. *Biomaterials, 34*(16), 3992–4001. https://doi.org/10.1016/j.biomaterials.2013.02.011.

Yap, K. K., Gerrand, Y.-W., Dingle, A. M., Yeoh, G. C., Morrison, W. A., & Mitchell, G. M. (2020). Liver sinusoidal endothelial cells promote the differentiation and survival of mouse vascularised hepatobiliary organoids. *Biomaterials, 251*. https://doi.org/10.1016/j.biomaterials.2020.120091, 120091.

Ye, S., Boeter, J. W. B., Mihajlovic, M., van Steenbeek, F. G., van Wolferen, M. E., Oosterhoff, L. A., et al. (2020). A chemically defined hydrogel for human liver organoid culture. *Advanced Functional Materials, 30*(48), 2000893. https://doi.org/10.1002/adfm.202000893.

Ye, S., Boeter, J. W. B., Penning, L. C., Spee, B., & Schneeberger, K. (2019). Hydrogels for liver tissue engineering. *Bioengineering (Basel), 6*(3). https://doi.org/10.3390/bioengineering6030059.

Yu, C., Ma, X., Zhu, W., Wang, P., Miller, K. L., Stupin, J., et al. (2019). Scanningless and continuous 3D bioprinting of human tissues with decellularized extracellular matrix. *Biomaterials, 194*, 1–13. https://doi.org/10.1016/j.biomaterials.2018.12.009.

Zeng, W.-Q., Zhang, J.-Q., Li, Y., Yang, K., Chen, Y.-P., & Liu, Z.-J. (2013). A new method to isolate and culture rat kupffer cells. *PLoS One, 8*(8), e70832. https://doi.org/10.1371/journal.pone.0070832.

Zhai, X., Wang, W., Dou, D., Ma, Y., Gang, D., Jiang, Z., et al. (2019). A novel technique to prepare a single cell suspension of isolated quiescent human hepatic stellate cells. *Scientific Reports*, *9*(1), 12757. https://doi.org/10.1038/s41598-019-49287-7.

Zhu, C., Coombe, D. R., Zheng, M. H., Yeoh, G. C., & Li, L. (2012). Liver progenitor cell interactions with the extracellular matrix. *Journal of Tissue Engineering and Regenerative Medicine*, *7*(10), 757–766. https://doi.org/10.1002/term.1470.

Zhu, M., Koibuchi, A., Ide, H., Morio, H., Shibuya, M., Kamiichi, A., et al. (2018). Development of a new conditionally immortalized human liver sinusoidal endothelial cells. *Biological & Pharmaceutical Bulletin*, *41*(3), 440–444. https://doi.org/10.1248/bpb.b17-00661.

Zigmond, E., Samia-Grinberg, S., Pasmanik-Chor, M., Brazowski, E., Shibolet, O., Halpern, Z., et al. (2014). Infiltrating monocyte-derived macrophages and resident kupffer cells display different ontogeny and functions in acute liver injury. *Journal of Immunology*, *193*(1), 344–353. https://doi.org/10.4049/jimmunol.1400574.

CHAPTER TWO

Organoid systems for recapitulating the intestinal stem cell niche and modeling disease *in vitro*

Hui Yi Grace Lim[a,†], Lana Kostic[a,†], and Nick Barker[a,b,c,]*

[a]A*STAR Institute of Molecular and Cell Biology, Singapore, Singapore
[b]Kanazawa University, Kanazawa, Japan
[c]Department of Physiology, Yong Loo Lin School of Medicine, National University of Singapore, Singapore, Singapore
*Corresponding author: e-mail address: nicholas_barker@imcb.a-star.edu.sg

Contents

1. Introduction   57
2. Intestinal stem cells and their *in vivo* niche   60
3. Modeling the ECM using synthetic matrices   63
4. Development of "mini-gut" organoid models   67
5. Organoids as a platform for modeling intestinal regeneration, disorders, and cancer   72
   5.1 Organoid models for studying intestinal epithelium damage response   73
   5.2 Exploring the intestinal response to pathogens using organoid systems   75
   5.3 Modeling intestinal disorders *in vitro*   76
   5.4 Colorectal cancer in a dish   78
6. Conclusions   83
Acknowledgments   84
References   84

1. Introduction

The intestinal epithelium is the most rapidly renewing tissue within the adult mammalian organism (van der Flier & Clevers, 2009). Apart from its digestive and absorptive roles, the intestine also forms a physical barrier against microorganisms and other foreign material present in the intestinal

---

[†] These authors contributed equally.

Advances in Stem Cells and their Niches, Volume 6
ISSN 2468-5097
https://doi.org/10.1016/bs.asn.2021.10.001

lumen. This harsh luminal environment necessitates rapidly cycling cells and the complete replenishment of the epithelium every 3–5 days (Barker, Bartfeld, & Clevers, 2010). Adult stem cells fuel this constant turnover of the epithelium, by generating the diversity of differentiated, postmitotic cell types needed for the maintenance of intestinal homeostasis and regulation of intestinal functions. Understanding how the intestinal stem cell compartment and its local niche orchestrate this highly dynamic process to ensure tissue homeostasis and repair is crucial for driving regenerative medicine advances and the development of clinical strategies against stem cell-driven intestinal diseases.

Morphologically, the simple columnar epithelial layer making up the intestine lining is composed of alternating villi and crypts (Fig. 1). Villi are the finger-like protrusions extending into the intestinal luminal space, whereas the adjacent crypts are deep invaginations into the tissue from the villus base. These latter crypts are hubs of proliferative activity, containing actively cycling stem cells and their immediate progenitors, the transit-amplifying cells (Clevers, 2013; van der Flier & Clevers, 2009). Transit-amplifying cells are able to undergo several rounds of cell division before acquiring one of four differentiated lineage fates: (1) Paneth cells, which secrete antibacterial peptides and lysozymes that regulate the gut microbial environment (Clevers & Bevins, 2013); (2) enterocytes, which function primarily in nutrient absorption and transport of antigens (Kong, Heel, McCauley, & Hall, 1998; Snoeck, Goddeeris, & Cox, 2005); (3) mucous-producing goblet cells (Birchenough, Johansson, Gustafsson, Bergström, & Hansson, 2015; Kim & Ho, 2010); and (4) enteroendocrine cells that secrete gut hormones (Gribble & Reimann, 2016). In addition, other minor cell types have also been identified, including microfold cells (M cells), cup cells, Peyer's patch-associated, and tuft cells, though their functions are currently not well defined (Debard, Sierro, & Kraehenbuhl, 1999; Gerbe, Legraverend, & Jay, 2012; Kyd & Cripps, 2008; McElroy, Frey, Torres, & Maheshwari, 2018). These differentiated cells migrate along the crypt-villus axis to occupy distinct positions, and most of them ultimately activate an apoptotic program at the villus tip where they become extruded into the intestinal lumen (Hall, Coates, Ansari, & Hopwood, 1994). An exception is Paneth cells, which together with intestinal stem cells are restricted to crypt bases and have an extended lifespan of around 30 days (Clevers & Bevins, 2013).

Here we present the major components of the *in vivo* intestinal stem cell niche and highlight the complementary *in vitro* systems that have since been

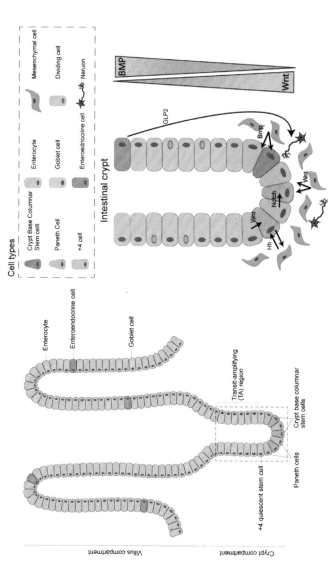

Fig. 1 Schematic illustration of the intestinal epithelium showing the crypt-villus morphology and the different cell populations present within each compartment. The key signaling pathways active within the crypt are also highlighted. Stromal/mesenchymal cells produce key signaling molecules that regulate the behavior of the epithelial cells within the niche to ensure stem cell maintenance and proliferation while inducing differentiation in cells migrating upward.

developed to model key aspects of this niche. We discuss the development of synthetic matrices and introduce various cell- and organoid-based models that recapitulate to some degree the molecular, cellular, and biophysical features of the native intestinal stem cell niche. To further extend the complexity of these *ex vivo* models, we discuss newer studies incorporating additional elements of the native microenvironment such as the immune system, nervous system, and gut microbiome. Finally, we consider the implications of utilizing these models to investigate stem cell-driven intestinal disorders, regeneration, and carcinogenesis.

2. Intestinal stem cells and their *in vivo* niche

Establishing a consensus on the identity and localization of the stem cell pool in the small intestine has long been an obstacle in stem cell research due to the lack of adequate markers and methods to validate their status. Initially, cells located at the +4 position were marked as potential stem cells, because of their ability to retain labels and prominent sensitivity to irradiation (Cairnie, Lamerton, & Steel, 1965; Potten, 1977). A new model was later proposed, identifying the columnar cells at the crypt base as potential adult stem cells (Cheng & Leblond, 1974). Technological advances over the ensuing decades have since facilitated the identification and functional validation of several markers labeling putative stem cell populations at both positions.

The first specific marker of crypt base stem cells, Lgr5, was defined in 2007 by Barker et al. (2007). Each crypt contains around 14 actively dividing Lgr5 + cells intermingled with Paneth cells (Snippert et al., 2010). Aided by the generation of inducible mouse models that allowed for visualization of specific cell populations and lineage tracing of their progeny, Lgr5 + cells were found to produce long-lived clones stretching from the crypt base to the tip of the villi (Barker et al., 2007). Additional confirmation of their stemness came from *in vitro* experiments, whereby isolated intestinal Lgr5 + cells displayed the ability to robustly generate organoids encompassing all the differentiated cell lineages, which could also be sustained in the long term (Barker et al., 2007; Sato et al., 2009). In addition to Lgr5, other markers expressed by cells in the crypt base include Rnf43, Znrf3, Olfm4, Ascl2, and Msi1 (Muñoz et al., 2012; Potten et al., 2003; van der Flier et al., 2009; van der Flier, Haegebarth, Stange, van de Wetering, & Clevers, 2009). An alternative stem cell marker, Bmi1, located at the +4 position was later proposed (Sangiorgi & Capecchi, 2008). Lineage tracing studies showed that Bmi1+ cells are distinct from Lgr5+ cells, and represent a relatively quiescent pool that becomes activated upon injury (Sangiorgi & Capecchi, 2008; Yan et al., 2012). Other markers also found to be expressed in the +4 cells include Lrig1, Hopx, and mTert (Itzkovitz et al., 2011; Powell et al., 2012; Takeda et al., 2011). However, several studies have also reported strong Bmi1 expression in Lgr5+ cells, as well as a broader zone of Bmi1 expression around the crypt (Itzkovitz et al., 2011; Muñoz et al., 2012; van der Flier, van Gijn, et al., 2009). These findings have made the use of Bmi1 as an exclusive stem cell marker controversial, as it appears that it is not uniquely expressed in +4 cells but broadly around the crypt.

Intestinal stem cells are regulated by a variety of chemical and physical signals originating from the surrounding epithelial crypt cells and pericryptal stroma that create a unique stem cell niche (Fig. 1). Exchange of information between the niche and epithelium is critical in maintaining the homeostatic cellular state, and also enables adequate cell response in cases of tissue trauma. One key component of the intestinal stem cell niche is crypt-base Paneth cells, which regulate the normal physiological state of the small intestine through the secretion of ligands such as Wnt3A, Notch, and EGF that directly influence Lgr5+ cells through modulation of key signaling cascades (Sato, Stange, et al., 2011; Sato, van Es, et al., 2011). Purified individual Paneth cells are incapable of forming organoids but co-culture of Lgr5+ and Paneth cells generates more organoids than Lgr5+ cells alone (Sato, Stange, et al., 2011; Sato, van Es, et al., 2011; Takeda et al., 2011), highlighting the importance of these epithelial niche signals for stem cell maintenance. After injury, Paneth cells can respond by modulating signaling pathways that result in increased proliferation and differentiation of crypt stem cells. Contrary to the classical definition of cell hierarchy, terminally differentiated cells often show significant plasticity, which allows them to undergo trans- or de-differentiation under pathological conditions. In conditions of trauma, Paneth cells may lose their characteristic markers, such as lysozyme, dedifferentiate and transition to a proliferative state to counter for the loss of cells in the crypt (Demitrack & Samuelson, 2016; Yu et al., 2018). Other physiological conditions, like starvation, can moderate the mTORC1 cascade in Paneth cells, which has an influence on the size of the Lgr5+ cell pool. Calory restriction reduces mTORC1 signaling only in Paneth cells and has no direct effect on intestinal stem cells. In response to this change in mTORC, Paneth cells modulate the ISC state toward increased self-renewal. Such a plastic response to calorie availability allows for preservation of the stem cell pool and enables better adaptation to periods of prolonged starvation (Yilmaz et al., 2012).

Another source of niche signals is the stromal cells surrounding the intestinal crypt. The *lamina propria*, which is situated directly underneath the epithelium, harbors mesenchymal cells, endothelial cells, and fibroblasts, with neural and inflammatory cells dispersed between them (Pinchuk, Mifflin, Saada, & Powell, 2010). These cells function not only as structural support for the crypt but also provide signals that regulate crypt-base columnar cell activity. Ablation of Foxl1 mesenchymal cells from the stromal compartment resulted in rapid attenuation in a proliferation of CBC and TA epithelial cells due to the loss of Wnt, highlighting the importance of the stromal compartment in intestinal maintenance (Aoki et al., 2016). Another key signaling molecule is Bone Morphogenic Protein (BMP), which is maintained at low levels in the intestinal crypt but at high levels in the niche compartment. Bmp antagonizes Wnt and protects against extensive amplification of the stem cell pool in the niche (Miyazono, Kamiya, & Morikawa, 2010; Qi et al., 2017). The signals are not, however, transferred unidirectionally from the stroma to the crypt. Epithelial cells also secrete Hedgehog ligands that are stimulating hedgehog signaling in the stroma, where they play a role in patterning of the gastrointestinal tract (Kolterud et al., 2009; Zacharias et al., 2011). Neural cells from the stroma, but not CBC cells, express the GLP-2 receptor. These stromal neurons respond to GLP2 presence and subsequently transfer the signal to the epithelium, which induces enterocyte progenitor proliferation (Bjerknes & Cheng, 2001). Finally, during injury response, modulation of YAP/TAZ signaling was shown to induce intestinal stem cell proliferation, but also the remodeling of extracellular matrix (ECM) and collagen I production, which stimulates epithelial cell fate change. In this condition, epithelial cells start expressing embryonal markers and enter a highly proliferative state that enhances healing. Importantly, this effect of YAP/TAZ on regenerative processes was recapitulated *in vitro*, providing the opportunity for modeling cell fate and reprogramming in culture systems (Imajo, Ebisuya, & Nishida, 2015; Yui et al., 2018). These observations highlight that cooperation between the epithelium and mesenchyme is not only critical for maintaining homeostasis, but also for effecting a proper response to external challenges and facilitating efficient tissue repair.

3. Modeling the ECM using synthetic matrices

In vitro cell-based assays have been a key pillar of biological research for decades. These models, particularly 2D monolayer cell cultures, have been widely adopted as they offer relatively simple and inexpensive solutions for a wide range of studies. However, the lack of a three-dimensional context means that such 2D systems continue to fall short of accurately replicating the *in vivo* environment and could induce significant alterations to native cellular behavior and response to stimuli. The development of 3D systems capable of mimicking near-physiological conditions *ex vivo* is therefore of vital importance for the study of cell biology, performing functional screens, preclinical studies, and the development of precision medicine.

The most commonly used 3D culture platforms are air-liquid interface (ALI) constructs, spheroids/organoids grown in various matrices, and organ-on-a-chip models (Fig. 2A) (Choi, Wu, Lee, Baquir, & Hancock, 2020). ALI systems typically utilize a permeable membrane that allows cells to be suspended within the culture medium while maintaining an air interface. Intestinal ALI constructs were successfully generated from fragments of epithelial and mesenchymal cells grown in a collagen gel with an air interface (Ootani et al., 2009), although these cells were unable to accurately recapitulate the characteristic crypt-villus intestinal morphology. This limitation was circumvented by the development of an organoid model by Sato et al., which incorporated a Matrigel matrix that supported the formation of 3D spheric structures harboring crypt-like and villus-like domains (Sato et al., 2009). Importantly, these intestinal organoids could be passaged long-term without accumulating genetic changes, establishing them as a near-physiological renewable source of intestinal epithelium for a variety of basic biology, pre-clinical screening, and precision medicine applications. 3D organoid culture derived from adult stem cells is, however, not able to fully replicate all the functions of a real organ. These systems are lacking stromal compartments and lamina propria, both known to play a critical role in the functioning of the intestine (Pinchuk et al., 2010; Powell, Pinchuk, Saada, Chen, & Mifflin, 2011). Organoid systems are also missing immune cells which decreases the significance of the organoid-derived results for drug testing and cancer therapies. Lack of vascularization that could limit organoid growth was partially countered by development of spinning

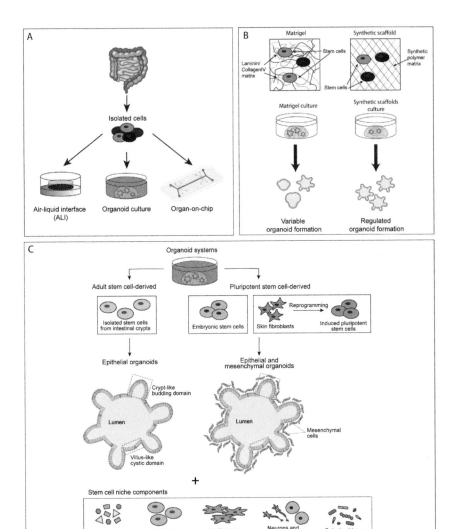

Fig. 2 Organoid models for recapitulating *in vivo* intestinal morphology and function. (A) Single intestinal stem cells or crypt fragments harboring these stem cells can be isolated from primary intestinal tissue. These cells can be maintained in 3D culture using three main approaches: Air–liquid interface (ALI), organoids embedded in scaffolds, and microfluidic organ-on-chip models. (B) Comparison of Matrigel and synthetic scaffolds for organoid culture. The complex and variable structure of Matrigel can result in heterogeneous organoid growth, unlike synthetic alternatives that can be precisely defined and adjusted to deliver standardized formation outcomes. (C) Organoid systems derived from adult or pluripotent stem cells. Adult stem cell-derived organoids harbor budding and cyst domains that reflect the native crypt-villus intestinal morphology. Organoids derived from pluripotent stem cells have an additional mesenchymal component surrounding the epithelial cell layer. Both organoid systems can be further combined with elements of the native intestinal stem cell niche, including signaling/growth factors, Paneth cells, myofibroblasts, neurons, immune cells, and the gut microbiome to reflect the complexity of interactions between these diverse components *in vivo*.

bioreactors that allow for media exchange and access to nutrients (Velasco, Shariati, & Esfandyarpour, 2020). Moreover, accurate replication of intestinal organ in a dish also needs a nervous system component. Successful attempts to generate colonic organoids in co-culture with enteric neurons were done with human embryonal cells (Park, Nguyen, & Yong, 2020). However, the system could not sustain simultaneous generation of vascularization with neurons and lacked immune system components. Another solution for incorporating the niche compartments *in vitro* includes combining of 2D and 3D culture in Transwell® membrane plates, creating a "lumen" and "lamina propria" in different sections of the well and providing a system for studying epithelial–immune interactions and response to pathogens (Noel et al., 2017). Moreover, 3D organoids can also be combined with microfluidics for development of *in vitro* organ-on-chip models. Microfluidic chips are miniature devices with chambers connected with microchannels that allow the controlled flow of fluids between them. Capturing near-physiological functions using organ-on-chip models has been achieved successfully for several organ systems, including the gastrointestinal tract (Chen, Zhang, Zhang, & Liu, 2020; Esch, Mahler, Stokol, & Shuler, 2014). This approach allows for the reconstruction of microenvironmental interactions between the tissues, similar to living organs, but the systems are often made to replicate only a specific feature of interest for each organ. Nevertheless, organ-on-chip technology also lacks vascularization, and remains incompletely adapted for high-throughput screening that is necessary for drug discovery studies (Fang & Eglen, 2017). With additional validation and optimization, this advanced technology has the potential to revolutionize therapy discovery.

The key factor across these models is the integration of an extracellular scaffold in the system to support 3D culture. The extracellular matrix (ECM) is a non-cellular component of tissues composed mostly of glycosaminoglycans and proteins (Frantz, Stewart, & Weaver, 2010). In living organisms, the ECM can undergo extensive remodeling that results in changes to its stiffness, density, and elasticity, parameters that contribute to proper tissue development and function (Engler, Sen, Sweeney, & Discher, 2006). Incorporating this ECM component to *in vitro* systems is thus important in order to faithfully replicate the physiology and function of the intestine. *In vitro*, the ECM plays an essential role in establishing proper tissue architecture and in providing mechanical cues that influence signaling pathways (Abdeen, Lee, & Kilian, 2016). This vital component allows generation of 3D organoid cultures, as it supports their growth and directs cell fate and behavior, by providing stiffness for the system allowing proper cell renewal, differentiation and organization, and creating a functional replicate of *in vivo* architecture. Matrices used for facilitating *in vitro* growth of 3D structures can be generated in a variety of ways, ranging from chemically defined synthetic materials to less well characterized ECM components extracted from living cells (Blow, 2009; Hughes, Postovit, & Lajoie, 2010).

This latter category of matrices includes Matrigel, which was developed more than 40 years ago from a murine Engelbreth-Holm-Swarm sarcoma tumor that produces a large amount of ECM *in vivo* (Orkin et al., 1977). The main structural components of Matrigel are laminin and collagen IV (~90%), combined with minor portions of enactin and perlecan (~10%) (Fig. 2B) (Aisenbrey & Murphy, 2020). Apart from ECM structural components, Matrigel is also rich in growth factors originating from tumor cells (Kleinman & Martin, 2005). These factors include epidermal growth factor (EGF), fibroblast growth factor (FGF), insulin-like growth factor 1 (IGF-1), and transforming growth factor-beta (TGF-β) (Talbot & Caperna, 2015). Moreover, the physical properties of Matrigel change with temperature: it is liquid at low temperatures and turns into a gelatinous substance at temperatures between 22 and 37 °C. This feature is a result of bonds created between laminin and collagen IV, via enacting crosslinking activity. Because of these properties, Matrigel has been widely used in research, gaining even more importance in the past decade with the development and expansion of organoid cultures. However, due to their sarcoma cell origin, the concentration of ECM components and growth factors can vary significantly between batches. Studies conducted on Matrigel to elucidate its composition detected around 14,000 peptides, with varying functions and a significant difference in their ratios between lots (Hughes et al., 2010; Kleinman et al., 1986). These inconsistencies in components can have a significant impact on Matrigel's mechanical and chemical properties, calling for special attention in certain assays. For example, differences in growth factor concentration can have an impact on growth assays. More importantly, tumor-derived Matrigel represents a significant obstacle for utilizing Matrigel-embedded organoids in human therapies.

In an attempt to overcome some of the limitations of Matrigel, several synthetic alternatives were developed over the past decades in order to offset the inconsistency and variability present between Matrigel batches, and to improve on safety for the ultimate use of organoids in humans. The main advantage of synthetic matrices is their ability to be precisely modified for a specific use, without altering any other property of the matrix (Li, Sun,

Li, Kawazoe, & Chen, 2018; Spicer, 2020). Synthetic polymers used for generating matrices come in a wide variety of forms, but some of the most commonly used scaffolds are derived from polyacrylamide (PAM) and polyethylene glycol (PEG). PAM is a synthetically produced polymer without a charge, which is inert to reactions with biological agents and able to form hydrogels. When combined with other synthetic components, PAM can interact with cells. PAM-based materials are widely used for cell-based assays, but they remain limited to 2D cell culture due to the toxicity of hydrogel precursors and the lack of proper cell encapsulation (Caliari & Burdick, 2016; Li et al., 2018). PEG is another widely used synthetic polymer with features similar to PAM. This polymer is also bioinert, forms hydrogels, and can be easily manipulated. By combining PEG with various reactive groups, its polymerization and hydrogel initiation properties can change. Polymerization of this matrix can occur via enzymatic reactions, resulting in the formation of non-toxic precursors and better cell encapsulation (Caliari & Burdick, 2016; Ehrbar et al., 2007). In other cases, hydrogel formation occurs via photopolymerization by exposing the polymer to visible or UV light (Fairbanks et al., 2009; Nguyen & West, 2002). Combination of PEG with maleimide groups resulted in PEG-4MAL, a synthetic hydrogel successfully utilized for *in vivo* delivery of human pluripotent stem cell (hPSC)-derived organoids to injured colon in mice (Cruz-Acuña et al., 2018). This indicates that 3D culture based on PEG can be considered for developing therapies that can be used in treating human diseases. However, before delivering organoid based therapies to humans, more work is needed to properly define the properties of synthetic polymers and their effect on the embedded cells, in terms of differentiation, proliferation and mutation rate.

4. Development of "mini-gut" organoid models

Intestinal organoids are *in vitro* three-dimensional models of the native intestinal epithelium. Unlike traditional monolayer cell culture systems with limited structural and cellular complexity, organoids display striking similarities to the *in vivo* epithelium, being composed of a villus-like cystic structure enclosing a central lumen and crypt-like budding domains (Rahmani, Breyner, Su, Verdu, & Didar, 2019) (Fig. 2C). These cysts and buds accommodate the Lgr5 + intestinal stem cells and all the differentiated cell lineages within specific locations and at numbers that closely approximate the *in vivo* situation (Sato et al., 2009). Although animal models indeed provide the cellular heterogeneity and diversity of interactions that could yield more physiologically relevant insights, organoid technology has emerged as a viable alternative due to its ability to effectively recapitulate molecular, cellular, and physical features of the native condition. This has enabled the study of key aspects of stem cell biology and visualization of epithelial dynamics *in vitro* in real-time. Moreover, the possibility of utilizing patient-derived samples to generate human intestinal organoids offers a valuable system to study key aspects of human intestinal biology.

Currently, two primary methods are used to establish intestinal organoids *in vitro*. The first involves the isolation of either adult stem cells from intestinal crypts or crypt fragments harboring live stem cells to form epithelial organoids that can be sustained and self-renewed over extensive periods of time. Despite the absence of stromal components in this organoid system, as previously discussed, the use of exogenous growth factors within the culture media combined with a Matrigel-based matrix allowed for the establishment of intestinal organoids from individual Lgr5 + stem cells for the first time (Sato et al., 2009). This method utilizes a combination of epidermal growth factor (EGF), R-spondin1, and Noggin, signaling molecules previously shown to play key roles in crypt proliferation and maintenance *in vivo* (Haramis et al., 2004; Kim et al., 2005; Pinto, Gregorieff, Begthel, & Clevers, 2003). A separate injury-inducible Bmi1 + intestinal stem cell population displayed a similar ability to form long-lived spheroids when isolated in culture, in the presence of a growth factor cocktail comprising EGF, Noggin, R-spondin1, and Jagged (Yan et al., 2012). Importantly, unlike earlier reports of intestinal primary cultures with short-term viability and/or lacking sufficient complexity in recapitulating the native crypt-villus morphology (Evans, Flint, Somers, Eyden, & Potten, 1992; Fukamachi, 1992; Whitehead, Demmler, Rockman, & Watson, 1999), these stem cell-derived organoids could be passaged for over 8 months while retaining key stem cell properties (Sato et al., 2009). Close similarities in transcriptional signatures between cultured intestinal organoids and whole intestinal crypts (Sato et al., 2009), and in gene expression profiles derived from region-specific intestinal organoid cultures and their corresponding crypt/villus preparations (Middendorp et al., 2014) have also been uncovered. Importantly, these region-specific intestinal organoids continue to retain their regional characteristics even after long-term *ex vivo* culture and following transplantation into immune-deficient mice (Middendorp et al., 2014; Tsai et al., 2017), a testament to the ability of such organoid cultures in retaining intrinsic molecular and physiological properties of their tissue of origin.

Robust growth of human intestinal organoids *in vitro* has also been established from patient biopsies and tissue samples. They take the form of cyst-like structures and can be sustained long-term with the addition of Wnt3a (Sato, Stange, et al., 2011; Sato, van Es, et al., 2011). Screening of a panel of small molecules further identified Alk inhibitor and p38 inhibitor as additional components promoting human intestinal organoid growth and facilitating the formation of crypt-like budding domains (Sato, Stange, et al., 2011, Sato, van Es, et al., 2011). Given that p38 is also important in driving differentiation along secretory lineages (Houde et al., 2001; Otsuka et al., 2010), the inclusion of p38 inhibitor in this culture media makes it primarily suitable for maintaining an undifferentiated, stem cell state within the organoids to sustain their growth rather than generating the differentiated lineages of the intestinal epithelium. To address this, insulin-like growth factor 1 (IGF-1) and fibroblast growth factor 2 (FGF-2) were subsequently identified as additional niche factors that are able to replace p38 (Fujii et al., 2018), promoting both sustained organoid growth and the diversity of differentiated cell lineages to recapitulate *in vivo* human intestinal homeostasis. Importantly, these human intestinal organoid models appear to retain their native intestinal functionality of facilitating ion transport. Like human intestinal tissues, the organoids express multiple apically and basolaterally localized transport proteins including NHE3, CFTR, and Na^+/K^+ ATPase in region-specific patterns depending on their origin from the duodenum, jejunum or ileum, and also display Na^+ absorption and Cl- secretion activities (Kovbasnjuk et al., 2013; Zachos et al., 2016).

A second method for the establishment of intestinal organoids relies on the use of embryonic stem cells (ESCs) or induced pluripotent stem cells (iPSCs). In this approach, pluripotent stem cells are first induced to form definitive endoderm using activin A, followed by FGF4- and Wnt3a-directed differentiation toward mid- and hind-gut lineages (Spence et al., 2011). This trajectory parallels the processes involved in embryonic intestinal development. Mid- and hind-gut spheroids are then cultured in Matrigel in the presence of R-spondin1, EGF, and Noggin to induce the formation of intestinal organoids with their characteristic villus-like and crypt-like cyst and bud domains (McCracken, Howell, Wells, & Spence, 2011; Spence et al., 2011). A crucial difference between PSC-derived and adult stem cell-derived intestinal organoids is the presence of a mesenchymal layer surrounding the epithelium in the former, which is able to differentiate into fibroblasts, smooth muscle cells, and myofibroblasts. These mesenchymal cells serve as important stem cell niche components *in vivo*, and their inclusion in this organoid system further supports transplantation of these PSC-derived organoids into immune-deficient mice to function as *in vivo* models of the human intestine (Watson et al., 2014), although more recently, PSC-derived intestinal organoids lacking mesenchymal cells have also been generated (Mithal et al., 2020). At the transcriptomic level, PSC-derived intestinal organoids display a fetal-like molecular signature that aligns more closely with the developing embryonic intestine rather than the adult tissue (Finkbeiner et al., 2015), limiting their relevance in modeling the mature adult intestine. Critically, the pluripotent nature of these cells also raises concerns of unrestricted growth and teratoma formation that pose a significant barrier for translation to the clinic (Hentze et al., 2009). Nevertheless, when transplanted into mice, the PSC-derived intestinal organoids turn on hallmarks of maturation, including elevated expression of the adult intestinal stem cell marker OLFM4 (Finkbeiner et al., 2015). Therefore, in spite of its fetal characteristics, PSC-derived organoids continue to represent a unique model replicating both epithelial and stromal components of the intestinal stem cell niche, as well as the fetal-to-adult developmental transition during the differentiation and maturation process.

While specific combinations of growth factors have proved sufficient to sustain robust intestinal organoid growth, other studies have explored the incorporation of cellular components to organoid cultures as an endogenous source of these signaling factors. A prime example is the Paneth cell, a key component of the intestinal stem cell niche physically adjacent to the intestinal stem cells that supplies niche signaling components Wnt3, EGF, and the Notch ligand Dll4 (Sato, Stange, et al., 2011; Sato, van Es, et al., 2011). Incorporating Paneth cells together with intestinal stem cells in culture significantly enhanced organoid formation, an effect that could be substituted by the direct addition of the Wnt3a ligand (Sato, Stange, et al., 2011, Sato, van Es, et al., 2011). Interestingly, depletion of Paneth cells *in vivo* did not result in major perturbations on intestinal crypt homeostasis (Durand et al., 2012; Garabedian, Roberts, McNevin, & Gordon, 1997), although other studies have reported alterations in stem cell numbers and proliferative capacity in response to Paneth cell loss (Bastide et al., 2007; Mori-Akiyama et al., 2007; Sato, Stange, et al., 2011; Sato, van Es, et al., 2011). Another possibility is the role of other components of the native intestinal stem cell niche, which could compensate for *in vivo* Paneth cell loss by supplying similar signaling molecules. This includes adjacent enteroendocrine and tuft cells (van Es et al., 2019), and underlying stromal cells surrounding the crypt base cells that also interact via Wnt, EGF, and Notch pathways (Farin, Van Es, & Clevers, 2012; Lahar et al., 2011). Therefore, Paneth cells are critical players in sustaining the intestinal stem cell niche both *in vivo* and *in vitro*, and function in tandem with a diversity of other supporting niche cells to supply signals directing stem cell maintenance and crypt homeostasis.

Co-culture organoid models have explored the inclusion of other stromal compartments to mimic the native intestinal stem cell niche, particularly for adult stem cell-derived organoids. The air-liquid interface culture system described earlier was one of the first developed models used to grow epithelial and stromal cells derived from neonatal intestinal tissue, which succeeded in sustaining long-term organoid growth (Ootani et al., 2009). A specific role for intestinal subepithelial myofibroblasts in supporting organoid growth *in vitro* was later found (Lahar et al., 2011; Pastuła et al., 2016). The myofibroblast-intestinal organoid co-cultures supported the growth of larger organoids as compared to monocultures, even in the absence of exogenous Rspo1 that is normally supplied as an essential growth factor to the organoid culture medium (Lei et al., 2014). In support of a role for myofibroblasts in modulating Wnt signaling, transcriptomic analysis confirmed elevated expression of several Wnt pathway modulators including Rspo2 and Dkk3 when compared against isolated non-supportive myofibroblasts (Lei et al., 2014).

The nervous and immune systems form yet another important component of the intestinal stem cell niche. Although *in vitro* models incorporating these cells remain limited, some progress has been made with the establishment of methods to direct human PSCs to differentiate into functional enteric neurons (Fattahi et al., 2016), and subsequently, to induce the concurrent development of enteric neurons with the growth of PSC-derived intestinal organoids *in vitro* (Workman et al., 2017). Moreover, co-cultures have been established between intestinal organoids and multiple immune cell populations, including intraepithelial lymphocytes (Nozaki et al., 2016), T lymphocytes (Jung et al., 2018), and innate lymphoid cells (Lindemans et al., 2015). Two-dimensional intestinal monolayers have also been generated from organoids to allow for greater accessibility to both the apical and basolateral membranes of the epithelium, for instance in a co-culture system incorporating macrophages (Noel et al., 2017). These studies have increasingly led to the identification of key signaling molecules and secreted cytokines involved in the immune response that modulate aspects of intestinal epithelial homeostasis. One such factor, interleukin-22 (IL-22), is produced by innate lymphoid cells following injury and was found to promote the growth and maintenance of the stem cell compartment within intestinal organoids, via STAT3 phosphorylation (Lindemans et al., 2015). Finally, the native intestinal epithelium is also persistently exposed to a diverse commensal gut flora that influences and regulates epithelial growth and function. Recent studies are attempting to recapitulate this relationship between epithelial cells and microbes *in vitro*, for example with the combined culture of *Lactobacillus* and intestinal organoids (Hou et al., 2018). These studies can be further extended into models of pathogenesis by incorporating co-cultures of pathogenic bacteria and viruses, which will be discussed in a later section. In all, the development of organoid models requires the integration of multiple cellular and molecular factors in order to recapitulate the complex network of interactions within the native intestinal stem cell niche.

5. Organoids as a platform for modeling intestinal regeneration, disorders, and cancer

The intestinal epithelium has tightly regulated mechanisms in place to ensure efficient daily tissue homeostasis and to effect rapid recovery after damage to maintain the integrity and function of this essential organ. Both the resident intestinal stem cells and their associated niche play critical roles in orchestrating this epithelial maintenance/repair. A detailed mechanistic understanding of stem cell-driven homeostatic renewal and regeneration in the intestinal epithelium is critically important for deciphering the underlying causes of intestinal disease and to facilitate the development of effective clinical treatments for conditions ranging from ulcerative colitis to Crohn's disease. Accurate models of epithelial homeostasis, regeneration and disease are essential tools for facilitating such mechanistic studies. Mouse models have undoubtedly proven useful for modeling homeostatic renewal and dysfunction in the gut, but it has proven challenging to develop more widely accessible, physiologically accurate *in vitro* models compatible with more high throughput functional studies/screens. Whilst transformed cell lines have provided important insights into the basic principles underlying epithelial function in the gut, a 2D system has many inherent limitations in accurately capturing and recapitulating the physiological and morphological properties of an organ. With the recent advances in 3D culture and organoid platform development, we are now much better placed to model various diseases and evaluate therapies in an *in vitro* system that more accurately mimics the complexity of real organs (Fig. 3). Such near-physiological

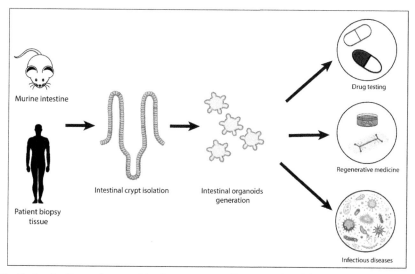

Fig. 3 Applications of organoid culture systems. Tissues isolated from murine organs or patient biopsies can be used to establish, expand and biobank *ex vivo* organoid cultures. These organoids can be utilized for drug discovery, modelling intestinal disorders and regenerative medicine applications using more advanced multiorgan tools such as organ-on-chip systems, and recapitulating host–microbe interactions by co-culturing organoids with pathogens.

model systems are readily amenable to a variety of functional screens, including acute injury response to cytotoxic agents such as bacteria and viruses, or therapeutics such as chemotherapeutics or gamma-irradiation (Jiminez, Uwiera, Douglas Inglis, & Uwiera, 2015; Lu et al., 2019; Yu, 2013).

5.1 Organoid models for studying intestinal epithelium damage response

The intestinal epithelium is relatively susceptible to acute injury resulting from a variety of drugs used in the treatment of common conditions in humans, including bacterial infection and cancer. Indeed, combinatorial cancer treatments incorporating different chemotherapeutics, cytotoxic drugs and irradiation commonly result in severe, potentially life-threatening side effects in the intestine, which are difficult to treat (Beck et al., 2004; Kumagai, Rahman, & Smith, 2018; Shadad, Sullivan, Martin, & Egan, 2013; Tao et al., 2020). The underlying basis of this impaired intestinal function and the ensuing regeneration following treatment cessation remains poorly understood. The idea of precision treatments has been particularly present in cancer research in the past decade, and there has been a strong effort in the field to establish reliable platforms for treatment response studies. Organoids have been successfully utilized for chemotherapy response prediction and confirmed the ability of 3D cultures to recapitulate histological and behavioral features of the tumor of origin (Brown et al., 2019; Ganesh et al., 2019). Similarly, organoids can be used for studying irradiation- or chemotherapy-induced stem cell regenerative response, either by isolating and growing intestinal organoids post-treatment or by generating healthy mouse or human organoids and exposing them to irradiation/chemotherapy *in vitro* (Ayyaz et al., 2019; Yamauchi et al., 2014).

A detailed understanding of intestinal stem cell responses within organoid culture models following exposure to exogenous agents is essential to support adoption of treatment regimes in the clinic. Tirado et al. showed that transplantation of Lgr5 + rectal stem cell–derived organoids was effective in ameliorating acute radiation damage *in vivo* following gamma-ray exposure (Tirado et al., 2021). Independently, radio-resistant Sox9 + cells were found to rapidly expand in irradiated intestinal organoids, recapitulating *in vivo* observations, and confirming the validity of the system for replicating endogenous cell behavior *in vitro* (Van Landeghem et al., 2012). Intestinal organoids irradiated with 6 Gy X-ray dose showed an increase in CD44 expression and overexpression of Wnt11, correlating with the decrease in Lgr5 (Liang et al., 2020), marking Wnt11 as another potential target for therapy development. A functional screen performed using human intestinal organoids found that deletion or inhibition of Puma, a p53 target gene, significantly reduces epithelial injury upon chemotherapy and irradiation therapy administration (Leibowitz et al., 2018). Suppression of negative Wnt pathway regulators, such as Blood Vessel Epicardial Substance (BVES/Popdc1), was reported to enhance organoid growth after the same cancer treatment (Reddy et al., 2016). To compensate for the absence of endogenous stem cell niche signals in the organoid systems, conditioned media from mesenchymal stem cells was used to treat irradiated organoids. Treated organoids displayed faster restoration of damaged crypts and increased overall survival (Kim, Han, et al., 2020; Kim, Kim, et al., 2020). The response was shown to be dependent on Wnt and Notch signaling, consistent with *in vivo* observations (Demitrack & Samuelson, 2016; Kim, Han, et al., 2020; Kim, Kim, et al., 2020), highlighting the utility of organoids for accurately modeling *in vivo* stress response physiology. Collectively, these studies underscore the potential of organoid models for faithfully recapitulating physiological responses in the intestine and for deciphering ISC regenerative responses as a prerequisite for the development of more effective treatments.

Although no treatment is currently approved for moderating radiation toxicity in the intestine, some studies have shown promising effects using different drug treatments on irradiated organoids. For example, the small molecule BCN057 was shown to induce proliferation of intestinal stem cells after radiation exposure in both human colon and mice jejunum organoids (Bhanja, Norris, Gupta-Saraf, Hoover, & Saha, 2018), and the anti-inflammatory agent auranofin delivered improved epithelial survival following irradiation of human tissue-derived intestinal organoids (Nag et al., 2019). These studies further highlight the potential of organoid models in the development of therapies for ameliorating irradiation side effects in cancer patients.

5.2 Exploring the intestinal response to pathogens using organoid systems

Apart from the damage caused by chemotherapeutics or radiation, the intestinal epithelium is also susceptible to damage by a variety of pathogens including *Yersinia, Salmonella, Citrobacter,* and *Cryptosporidium* (Sicard, Le Bihan, Vogeleer, Jacques, & Harel, 2017; Zhang et al., 2016). Mouse models have played a central role in helping to decipher the behavior and response of the epithelium to infection in the intestine (Martinez Rodriguez et al., 2012). Development of the intestinal organoid system has, for the first time, facilitated similar modeling of epithelial injury responses to infection in the human intestine, providing clinically important insights. Human intestinal organoids have been successfully infected with a range of viruses, including enterovirus, adenovirus, rotavirus and coronavirus, eliciting similar damage responses to those observed *in vivo* (Drummond et al., 2017; Holly & Smith, 2018; Lamers et al., 2020; Saxena et al., 2016). Organoid-based experiments revealed that enteroviruses replicate within enterocytes and, together with rotaviruses, disrupt tight junctions to cause loss of intestinal barrier function (Drummond et al., 2017), while human adenovirus infection demonstrated a previously unknown affinity of the virus for goblet cells in organoids (Holly & Smith, 2018). Intestinal stem cells in organoids were particularly sensitive to avian influenza virus, which affected their ability to proliferate and differentiate, resulting in a loss of Paneth cells (Huang, Hou, Ye, Yang, & Yu, 2017). These pioneering experiments demonstrate the potential of organoid platforms for enhancing our understanding of organ response to viral damage and providing clinically relevant insights into the effect of infections on stem cells and their niche.

The same model systems can also be used for studying epithelial response to bacterial or parasitic infection in the intestine. When infected with

Clostridium difficile, murine and human iPSC-derived intestinal organoids displayed changes in Paneth cell activity and frequency via modulation of Cytokine-JAK-STAT signaling activity in the crypt, preventing the differentiation of Lgr5 + cells toward the niche lineages (Liu et al., 2019). Apart from studying injury response, intestinal organoids are giving us for the first time an accurate insight into symbiotic relationships inside the GI tract. Lgr5 + cells, Paneth cells, and goblet cells in murine organoids increased their population size in response to secretions from *Bifidobacterium* and *Lactobacillus* spp. (Lee et al., 2018). Moreover, crypts isolated from mice without gut microbiota failed to generate organoids *in vitro* and regained that ability upon introduction of microbes to the culture. This is an indication that the regenerative capacity of the intestinal epithelium can be partially dependent on the presence of the microbiome, and that this interaction can be modeled *in vitro* (Zaborin et al., 2017, 2020).

5.3 Modeling intestinal disorders *in vitro*

Intestinal disorders are an increasingly common spectrum of conditions characterized by various combinations of symptoms related to the GI tract. The lower parts of the small intestine and the colon can be affected by ulcerative colitis (UC) and Crohn's disease, two forms of inflammatory bowel disease (IBD) (Kaser, Zeissig, & Blumberg, 2010). Inflammation-induced UC is caused by structural changes to the tight junctions, resulting in disruption of the epithelial barrier. Due to continuous erosion of the tissue from increased apoptosis, this condition leads to impaired absorptive transport processes and elevation of claudin-2 expression (Gitter, Wullstein, Fromm, & Schulzke, 2001; Schmitz et al., 1999). Crohn's disease is also caused by reductions to the abundance and complexity of tight junction proteins, negatively impacting the integrity of the epithelial barrier (Zeissig et al., 2004). Recent advances in treating IBD have focused on suppressing underlying chronic inflammatory processes by targeting pro-inflammatory cytokines, such as tumor necrosis factor (TNF)-beta (Nielsen & Ainsworth, 2013; Sandborn & Hanauer, 1999), interleukin (IL)-23 and IL-12 (Aggeletopoulou, Assimakopoulos, Konstantakis, & Triantos, 2018; Punkenburg et al., 2016). However, such treatments are largely restricted to providing acute symptom relief, without achieving the rapid and sustained repair of the damaged intestinal epithelium that is necessary to prevent the future development of ulcers and acquisition of neoplastic changes driving cancer development in IBD patients (Grivennikov, 2013;

Lichtenstein & Rutgeerts, 2010). Such limitations, combined with the fact that a significant proportion of patients are unresponsive to current therapies, highlights the urgent need to develop more effective treatments to restore epithelial barrier function in IBD patients. Recent findings, based on mRNA and protein level comparison in healthy and UC affected individuals, suggest that ECM degradation combined with neutrophil maturation plays an important role in UC development. Degraded ECM subsequently leads to fibrosis development rather than UC pathological condition (Kirov et al., 2019). Crohn's disease has also been associated with aberrant ECM remodeling, caused by KIAA199 enzyme that degrades glycosaminoglycan hyaluronan (Soroosh et al., 2016). This indicates that components of ECM can be potential biomarkers or therapeutic targets for IBD.

The cultured intestinal epithelium has therapeutic value as a source of healthy tissue for use in treating intestinal disease *in vivo*. As proof of principle, mouse intestinal organoids introduced to ulcerated colon in a DSS-induced colitis model were shown to permanently engraft and repair the damaged colonic epithelium, contributing both stem and differentiated cell lineages to the regenerated epithelial lining (Yui et al., 2012). A subsequent study demonstrated the ability of human intestinal organoids to stably integrate into the mucosa of immunodeficient mice and reconstruct human epithelium *in vivo* (Sugimoto et al., 2018). Such models utilizing xenograft organoid transplants will be invaluable for deciphering human tissue functions and for driving the development of novel therapies for chronic, refractive intestinal disease.

The utility of intestinal organoids for repairing injured mucosa *in vivo* can be a feasible solution in the future for treating intestinal disorders. Organoids can be derived from a patient's healthy tissue and, after *in vitro* expansion, engrafted at the lesion site, or subjected to genetic manipulation for correcting defects prior to transplantation. This approach was successfully demonstrated using CRISPR-edited cystic fibrosis patient-derived intestinal organoids (Geurts et al., 2020). However, organoid-based therapies for intestinal disorders are still in their early stages. In particular, the long-term genetic and epigenetic stability of cultured organoids must be definitively established to assuage safety concerns regarding therapeutic applications. It is also likely that epithelial organoids would demonstrate only limited clinical efficacy in treating impaired gut functions that extend beyond the epithelial layer. Development of more complex stem cell-derived organoids incorporating mesenchymal components capable of contributing to damaged stromal/muscle compartments would potentially circumvent this limitation. Use of PSC-derived organoids harboring non-epithelial intestinal tissues would be a potential alternative source of healthy tissue for transplantation. However, the embryonic characteristics displayed by current PSC organoid models raises significant safety concerns about their long-term stability. Although transplanted organoids in mice have so far not induced tumor formation (Yui et al., 2012), additional attention is needed for ensuring the safety of the treatment and estimating the rate of mutation accumulation during their *in vitro* propagation. The development of organoid-based therapies for intestinal disorders could potentially solve the problem of epithelial healing and improve patient prognosis, and in parallel reduce the risk of inflammation-induced pre-cancerogenic/ carcinogenic transformations.

5.4 Colorectal cancer in a dish

Colorectal cancer is one of the most prevalent cancers worldwide, with the third-highest number of cancer-linked deaths each year (Rawla, Sunkara, & Barsouk, 2019). Treatment options are currently limited to surgery for those presenting early stages of the disease and chemotherapy for patients with more advanced, metastatic tumors. However, with as many as a third of patients showing cancer recurrence, there is an urgent need to develop newer, more effective cancer therapies and diagnostic tools for early detection. This can be achieved by establishing *in vitro* models of colorectal cancer that recapitulate the different stages of cancer progression, providing a platform for high-throughput screens of potential drugs. These models also lend key insights into cancer biology through the dissection of dysregulated signaling pathways and niche conditions driving tumorigenesis, all of which can serve as potential therapeutic targets. Nevertheless, establishing accurate models that closely recapitulate features of the tumor and tumor microenvironment remains a central challenge, given the presence of intra-tumor heterogeneity, multiple colorectal cancer subtypes, and stages of disease progression, and difficulties in incorporating elements of the tumor microenvironment *in vitro*.

Human colorectal cancer cell lines are one of the most widely used *in vitro* models due to their accessibility and speed. These cells are typically grown as two-dimensional monolayers, and broadly display molecular signatures found in primary colorectal tumors, including mutations in Wnt, MAPK, and PI3K pathways (Ahmed et al., 2013; Kim, Han, et al., 2020;

Kim, Kim, et al., 2020; Mouradov et al., 2014). Cell line panels are also utilized as a better representation of the diversity of cancer subtypes. Nevertheless, individual cell lines, when compared to the parental tumor they originated from, can show significant deviations in molecular and phenotypic traits (Domcke, Sinha, Levine, Sander, & Schultz, 2013; Ertel, Verghese, Byers, Ochs, & Tozeren, 2006), likely due to the selective culture conditions they are grown under and the absence of structural and cellular components normally present in the native tumor microenvironment.

To more closely replicate the three-dimensional tumor architecture, other groups have established methods promoting the formation of colorectal cancer spheroids or organoids from primary tissues. This can be achieved using tissue derived from genetic mouse models of intestinal cancer harboring mutations commonly found in human colorectal cancers, such as the $APC^{Min/+}$ line (Moser, Pitot, & Dove, 1990), and the Lgr5-EGFP-ires-creERT2/$APC^{fl/fl}$ line displaying dysregulated Wnt signaling originating from the Lgr5+ stem cell compartment (Schepers et al., 2012). Organoids established from APC-deficient mouse intestinal adenomas formed cyst-like structures in culture, and displayed robust growth in the absence of R-spondin and Noggin (Sato, Stange, et al., 2011; Sato, van Es, et al., 2011). Similar methods have been used to establish organoid outgrowth from primary human colorectal cancer samples, via the isolation and purification of intestinal crypts (Sato, Stange, et al., 2011, Sato, van Es, et al., 2011), cell clusters (Kondo et al., 2011), or sorted cell populations with stem-like properties (Ricci-Vitiani et al., 2007). In line with hyperactivated Wnt signaling being a prominent feature of human colorectal tumors, culture conditions lacking R-spondin are capable of sustaining the growth of human colorectal cancer organoids (Sato, Stange, et al., 2011, Sato, van Es, et al., 2011). These cancer organoids are also hyperproliferative, thriving even in the absence of growth factors Noggin and EGF that typically sustain stem cell maintenance and proliferation in normal intestinal organoids (Sato, Stange, et al., 2011, Sato, van Es, et al., 2011). Together, the establishment of cancer organoid systems and the evaluation of factors necessary for sustaining organoid growth can inform us on the niche signaling pathways that regulate tumorigenesis.

Beyond the study of tumorigenic pathways, colorectal cancer organoids also support the evaluation and characterization of putative cancer stem cell populations that play critical roles in driving tumor growth and recurrence. Cancer stem cells make up a subpopulation of cells within the tumor load that exhibit stem cell properties of self-renewal and the ability to give rise to differentiated lineages of the tumor (Clevers, 2011). Importantly, they are thought to be more resistant to standard chemotherapy treatments and thus could be responsible for sustaining the longer-term persistence of cancerous cells even if the majority of the tumor bulk is initially eliminated (Clevers, 2011). Identification of markers labeling these cancer stem cell populations is therefore of significant clinical interest due to the possibility of developing novel targeted approaches against cancer stem cells that can address this issue of cancer recurrence in patients. For colorectal cancer, one of the earliest characterized cancer stem cell markers is CD133. CD133 + colorectal cancer cells consistently established tumors in immunodeficient mice more efficiently than CD133- cells (O'Brien, Pollett, Gallinger, & Dick, 2007; Ricci-Vitiani et al., 2007), and could be sustained in long-term culture as cancer spheroids while still maintaining its tumorigenicity (Ricci-Vitiani et al., 2007; Todaro et al., 2007). Subsequent studies went on to identify a suite of putative colorectal cancer stem cell markers, including CD44 (Dalerba et al., 2007), CD44v6 (Todaro et al., 2014), CD26 (Pang et al., 2010), ALDH1 (Huang et al., 2009) and Lgr5 (Kemper et al., 2012). Although the cancer stem cell potential of these cell populations is routinely evaluated by their tumor initiation capability in transplantation assays, *ex vivo* organoid cultures offer an alternative system that can be manipulated with greater ease and speed. For instance, isolated $Lgr5^{high}$ colorectal cancer cells demonstrated the ability to generate organoids *in vitro* more efficiently than their $Lgr5^{low}$ and $Lgr5^{neg}$ counterparts, indicative of their enhanced stem cell potential (Sato, Stange, et al., 2011, Sato, van Es, et al., 2011). Moreover, colorectal cancer organoids generated from $Lgr5^{high}$ cells could be genetically modified by CRISPR-Cas9 to incorporate a Cre knock-in allele at the Lgr5 locus for tamoxifen-inducible recombination of reporter alleles, allowing for visualization of lineage tracing dynamics within these organoid-derived colorectal cancer tumors (Shimokawa et al., 2017). As a further readout for the ability of these colorectal cancer stem cells to sustain the growth of the differentiated tumor bulk, knock-in of a Lgr5-iCaspase9 cassette within colorectal cancer organoids to specifically ablate the Lgr5 + cancer stem cells was able to efficiently diminish the tumor load in the short term (Shimokawa et al., 2017). This Lgr5 + population present in intestinal tumor organoids is also necessary for efficient liver metastasis establishment, as shown in a Lgr5-DTR-eGFP model that similarly targets Lgr5-expressing cells upon Diphtheria Toxin (DT) administration (de Sousa e Melo et al., 2017). In all, the integration of *in vitro* organoid culture techniques with efficient genetic manipulation to selectively target cell populations of interest has facilitated the identification and characterization of novel markers of cancer stem cells, with the potential to serve as clinically relevant drug targets for the treatment of colorectal cancer.

The ability to generate organoids from patient-derived tumor samples has further opened up the possibility of establishing human organoid libraries or biobanks (Fujii et al., 2016; van de Wetering et al., 2015). These organoid libraries, which can include a large diversity of tumor subtypes along with matched healthy-cancer pairs, are valuable resources for therapeutic screens and drug development. Moreover, patient-derived tumor organoids can be used for assessment of drug responses *in vitro*, which can ultimately help doctors tailor personalized treatment options to individual patients. This has been done for colorectal cancer organoids with oncogenic KRAS mutations to evaluate the effectiveness of RAS inhibitors and other drug combinations currently in use in the clinic (Verissimo et al., 2016). While it is also possible to culture primary tumor tissue for short time periods (Failli et al., 2009), or to perform transplantation of patient-derived samples into immunocompromised mice (Tentler et al., 2012) to evaluate drug response, these methods typically fall short of the scale and speed achievable in organoid models in order to perform high-throughput drug screening.

Organoid models established using primary tumor tissues can closely replicate features of the tissues that were derived from, but they often represent a single, late time point in the development of the tumor. Such models thus remain limited in showing the key alterations occurring during earlier stages of tumorigenesis, or in providing an understanding of the temporal changes during the progression of the tumor. This is especially relevant in the context of colorectal cancer, which develops from a well-characterized adenoma-carcinoma sequence of mutations (Fearon & Vogelstein, 1990). To recapitulate these mutational changes, CRISPR/Cas9 technology was used to perform site-directed mutation for APC in healthy intestinal cells, as Apc is typically the first gene to be altered during colorectal cancer progression (Schwank et al., 2013). Much like organoids derived directly from APC-deficient intestinal tissue, these APC-mutant intestinal cells generated cyst-like organoids in the absence of R-spondin1. More recently, two independent groups successfully introduced sequential mutations into healthy human intestinal organoids, generating quadruple (Drost et al., 2015) and quintuple (Matano et al., 2015) mutant organoids. Both studies selected for the mutant organoids by withdrawal of individual growth factors and showed that their multiple-mutant organoids were ultimately capable of growth in the absence of all niche factors. Moreover, when transplanted into immunodeficient mice, the AKSTP (APC, KRAS, SMAD4, TP53, and PI3KCA) quintuple mutant organoids established non-invasive tumors reminiscent of low-grade adenocarcinoma (Matano et al., 2015). Conversely, the quadruple mutants (APC, TP53, KRAS and SMAD4) seeded into nude mice grew into invasive carcinomas (Drost et al., 2015). A similar approach was subsequently used to generate a model of serrated colorectal cancer, a subtype present in a quarter of colorectal cancer patients that follows a unique mutational sequence and displays distinct molecular features from the major subtype. Here, the authors performed stepwise introduction of the $BRAF^{V600E}$ mutation to induce MAPK pathway hyperactivation, followed by mutations in Tgfbr2 that perturbs the TGF-b pathway, two Wnt pathway suppressors Rnf43 and Znrf3, the tumor suppressor p16Ink4a, and a regulator of DNA damage, Mlh1 (Lannagan et al., 2019). Just like earlier models of conventional colorectal cancer, organoids generated from this sequence of mutations grew out in the absence of niche factors and could establish invasive carcinomas when transplanted into nude mice. In all, organoid models of colorectal cancer incorporating specific combinations of oncogenic mutations can shed light into the dynamic alterations occurring during tumorigenesis and provide valuable platforms for therapeutic development and generation of novel drug targets.

The growth of a tumor is also closely supported by a complex array of extracellular matrix proteins and stromal cells, including tumor-associated fibroblasts, which together make up the tumor microenvironment. Changes in the tumor microenvironment have been shown to bolster cancer progression and could provide the signals needed to sustain tumor growth and recurrence even in the presence of anti-cancer drugs (Albini & Sporn, 2007). Therefore, models of colorectal cancer *in vitro* also require an accurate depiction of tumor microenvironment features and signals to simulate the interactions supporting carcinogenesis. Some studies have incorporated collagen-based hydrogels to replicate the native collagen–rich extracellular matrix, demonstrating that the organization of collagen fibers can influence the acquisition of an epithelial or mesenchymal phenotype in colorectal cancer cells *in vitro* (Devarasetty et al., 2020). Others have used three-dimensional co-culture systems to introduce stromal elements into the tumor cell cultures, such as connective tissue (Nyga, Loizidou, Emberton, & Cheema, 2013) and cancer-associated fibroblasts (Dolznig et al., 2011), as well as particular species of gut microbiota specifically enriched within the tumor microenvironment (Kasper et al., 2020). Interestingly, recent work has further illustrated how tumor cells can actively alter their microenvironment to favor the propagation of these cancerous cells at the expense of other healthy intestinal cells. When co-cultured with Apc-deleted mutant intestinal cells or in media derived from these cells, wildtype intestinal cells showed diminished organoid formation capacity (Flanagan et al., 2021; van Neerven et al., 2021). These mutant intestinal cells were found to secrete factors that promote BMP signaling and inhibit Wnt signaling (Yum et al., 2021) including a specific Wnt inhibitor NOTUM (Flanagan et al., 2021; van Neerven et al., 2021), which represses stem cell maintenance and instead drives wildtype stem cells to differentiate. In contrast, mutant cells are unaffected due to their intrinsic Apc mutation that enhances Wnt signaling downstream of NOTUM activity. As more complex *in vitro* models integrating multiple features of colorectal cancers and their microenvironment continue to be developed, further inroads into cancer biology and therapeutics can be made to deliver better diagnostic and treatment outcomes for patients.

6. Conclusions

New 3D models for recapitulating intestinal morphology and function have brought remarkable advances to the stem cell field. With the expansion of techniques for *in vitro* culture, we are in a position to more accurately recapitulate the intestinal morphology in 3D and gain better insights into how intestinal cells interact with each other, which signaling cascades are present in homeostasis, and how cells respond to tissue damage. Organoid studies have revealed not only previously unknown interactions between the intestinal epithelium and the gut microbiome, but also the response of the epithelial cells to bacterial and viral infections. The diversity of approaches available for organoid culture and the availability of cell- or synthetically derived matrices have also brought us a step closer toward utilizing these newly developed tools for advancing human therapies. Efforts are now focused on creating suitable platforms for drug screening and the development of personalized treatments for cancer patients. New gene-editing tools are also providing more options for engineering organoids with desired properties and allowing for safer and more controlled future applications. Nevertheless, translating findings from 3D models to clinically viable solutions still represents a great challenge. There are a number of hurdles to be overcome before these methods can be safely applied for human therapies.

An imperative for functional assays, drug screening and clinical applications is the development of more complex systems that encompass more than just the epithelial compartment. Some advances have been made with the PSC-derived organoids; however, they are still lacking critical immune components and vascularization. Another concern for the utilization of PSC-derived organoids over the adult tissue-derived ones is their retention of fetal characteristics that could promote uncontrolled growth upon transplantation. To capture the full potential of organoid systems, it is necessary to develop more complex systems, recapitulating not only the biology of the epithelium but also of the endogenous niche. In parallel, there is a need for the development of safe, well-defined, easy-to manipulate and cheaper synthetic matrices that can support organoid growth as efficiently as Matrigel. Before advancing any of these methods to the clinic, a rigorous evaluation of the long term genetic/epigenetic stability of human organoids is necessary in order to guarantee the safety and consistency of this system. In all, these advances will allow us to fully capture the versatility and efficacy of organoid models in recapitulating the native intestinal stem cell niche and illuminating key insights into intestinal regeneration, disease and cancer.

Acknowledgments

This work was supported and funded by the Agency for Science, Technology and Research (A*STAR), the Singapore Ministry of Health's National Medical Research Council Open-Fund Individual Research Grant (Grant no. OFIRG19may-0069), and the National Research Foundation Investigatorship (Award no. NRF12017-03). Illustrations were designed by pch.vector/Freepik and Vecteezy.com.

References

Abdeen, A. A., Lee, J., & Kilian, K. A. (2016). Capturing extracellular matrix properties in vitro: Microengineering materials to decipher cell and tissue level processes. *Experimental Biology and Medicine (Maywood, N.J.)*, *241*(9), 930–938. https://doi.org/10.1177/1535370216644532.

Aggeletopoulou, I., Assimakopoulos, S. F., Konstantakis, C., & Triantos, C. (2018). Interleukin 12/interleukin 23 pathway: Biological basis and therapeutic effect in patients with Crohn's disease. *World Journal of Gastroenterology*, *24*(36), 4093–4103. https://doi.org/10.3748/wjg.v24.i36.4093.

Ahmed, D., Eide, P. W., Eilertsen, I. A., Danielsen, S. A., Eknæs, M., Hektoen, M., et al. (2013). Epigenetic and genetic features of 24 colon cancer cell lines. *Oncogenesis*, *2*(9), e71. https://doi.org/10.1038/oncsis.2013.35.

Aisenbrey, E. A., & Murphy, W. L. (2020). Synthetic alternatives to matrigel. *Nature Reviews Materials*, *5*(7), 539–551. https://doi.org/10.1038/s41578-020-0199-8.

Albini, A., & Sporn, M. B. (2007). The tumour microenvironment as a target for chemoprevention. *Nature Reviews Cancer*, *7*(2), 139–147. https://doi.org/10.1038/nrc2067.

Aoki, R., Shoshkes-Carmel, M., Gao, N., Shin, S., May, C. L., Golson, M. L., et al. (2016). Foxl1-expressing mesenchymal cells constitute the intestinal stem cell Niche. *Cellular and Molecular Gastroenterology and Hepatology*, *2*(2), 175–188. https://doi.org/10.1016/j.jcmgh.2015.12.004.

Ayyaz, A., Kumar, S., Sangiorgi, B., Ghoshal, B., Gosio, J., Ouladan, S., et al. (2019). Single-cell transcriptomes of the regenerating intestine reveal a revival stem cell. *Nature, 569*(7754), 121–125. https://doi.org/10.1038/s41586-019-1154-y.

Barker, N., Bartfeld, S., & Clevers, H. (2010). Tissue-resident adult stem cell populations of rapidly self-renewing organs. *Cell Stem Cell, 7*(6), 656–670. https://doi.org/10.1016/j.stem.2010.11.016.

Barker, N., van Es, J. H., Kuipers, J., Kujala, P., van den Born, M., Cozijnsen, M., et al. (2007). Identification of stem cells in small intestine and colon by marker gene Lgr5. *Nature, 449*(7165), 1003–1007. https://doi.org/10.1038/nature06196.

Bastide, P., Darido, C., Pannequin, J., Kist, R., Robine, S., Marty-Double, C., et al. (2007). Sox9 regulates cell proliferation and is required for Paneth cell differentiation in the intestinal epithelium. *Journal of Cell Biology, 178*(4), 635–648. https://doi.org/10.1083/jcb.200704152.

Beck, P. L., Wong, J. F., Li, Y., Swaminathan, S., Xavier, R. J., Devaney, K. L., et al. (2004). Chemotherapy- and radiotherapy-induced intestinal damage is regulated by intestinal trefoil factor. *Gastroenterology, 126*(3), 796–808. https://doi.org/10.1053/j.gastro.2003.12.004.

Bhanja, P., Norris, A., Gupta-Saraf, P., Hoover, A., & Saha, S. (2018). BCN057 induces intestinal stem cell repair and mitigates radiation-induced intestinal injury. *Stem Cell Research & Therapy, 9*(1), 26. https://doi.org/10.1186/s13287-017-0763-3.

Birchenough, G. M. H., Johansson, M. E., Gustafsson, J. K., Bergström, J. H., & Hansson, G. C. (2015). New developments in goblet cell mucus secretion and function. *Mucosal Immunology, 8*(4), 712–719. https://doi.org/10.1038/mi.2015.32.

Bjerknes, M., & Cheng, H. (2001). Modulation of specific intestinal epithelial progenitors by enteric neurons. *Proceedings of the National Academy of Sciences of the United States of America, 98*(22), 12497–12502. https://doi.org/10.1073/pnas.211278098.

Blow, N. (2009). Cell culture: Building a better matrix. *Nature Methods, 6*(8), 619–622. https://doi.org/10.1038/nmeth0809-619.

Brown, S. L., Kolozsvary, A., Isrow, D. M., Al Feghali, K., Lapanowski, K., Jenrow, K. A., et al. (2019). A novel mechanism of high dose radiation sensitization by metformin. *Frontiers in Oncology, 9*, 247. https://doi.org/10.3389/fonc.2019.00247.

Cairnie, A. B., Lamerton, L. F., & Steel, G. G. (1965). Cell proliferation studies in the intestinal epithelium of the rat: I. Determination of the kinetic parameters. *Experimental Cell Research, 39*(2), 528–538. https://doi.org/10.1016/0014-4827(65)90055-8.

Caliari, S. R., & Burdick, J. A. (2016). A practical guide to hydrogels for cell culture. *Nature Methods, 13*(5), 405–414. https://doi.org/10.1038/nmeth.3839.

Chen, X., Zhang, Y. S., Zhang, X., & Liu, C. (2020). Organ-on-a-chip platforms for accelerating the evaluation of nanomedicine. *Bioactive Materials, 6*(4), 1012–1027. https://doi.org/10.1016/j.bioactmat.2020.09.022.

Cheng, H., & Leblond, C. P. (1974). Origin, differentiation and renewal of the four main epithelial cell types in the mouse small intestine. V. Unitarian Theory of the origin of the four epithelial cell types. *The American Journal of Anatomy, 141*(4), 537–561. https://doi.org/10.1002/aja.1001410407.

Choi, K.-Y. G., Wu, B. C., Lee, A. H.-Y., Baquir, B., & Hancock, R. E. W. (2020). Utilizing organoid and air-liquid interface models as a screening method in the development of new host defense peptides. *Frontiers in Cellular and Infection Microbiology, 10*(228). https://doi.org/10.3389/fcimb.2020.00228.

Clevers, H. (2011). The cancer stem cell: Premises, promises and challenges. *Nature Medicine, 17*(3), 313–319. https://doi.org/10.1038/nm.2304.

Clevers, H. (2013). The intestinal crypt, a prototype stem cell compartment. *Cell, 154*(2), 274–284. https://doi.org/10.1016/j.cell.2013.07.004.

Clevers, H. C., & Bevins, C. L. (2013). Paneth cells: Maestros of the small intestinal crypts. *Annual Review of Physiology, 75*(1), 289–311. https://doi.org/10.1146/annurev-physiol-030212-183744.

Cruz-Acuña, R., Quirós, M., Huang, S., Siuda, D., Spence, J. R., Nusrat, A., et al. (2018). PEG-4MAL hydrogels for human organoid generation, culture, and in vivo delivery. *Nature Protocols*, *13*(9), 2102–2119. https://doi.org/10.1038/s41596-018-0036-3.

Dalerba, P., Dylla, S. J., Park, I. K., Liu, R., Wang, X., Cho, R. W., et al. (2007). Phenotypic characterization of human colorectal cancer stem cells. *Proceedings of the National Academy of Sciences of the United States of America*, *104*(24), 10158–10163. https://doi.org/10.1073/pnas.0703478104.

de Sousa e Melo, F., Kurtova, A. V., Harnoss, J. M., Kljavin, N., Hoeck, J. D., Hung, J., et al. (2017). A distinct role for Lgr5(+) stem cells in primary and metastatic colon cancer. *Nature*, *543*(7647), 676–680. https://doi.org/10.1038/nature21713.

Debard, N., Sierro, F., & Kraehenbuhl, J. P. (1999). Development of Peyer's patches, follicle-associated epithelium and M cell: Lessons from immunodeficient and knockout mice. *Seminars in Immunology*, *11*(3), 183–191. https://doi.org/10.1006/smim.1999.0174.

Demitrack, E. S., & Samuelson, L. C. (2016). Notch regulation of gastrointestinal stem cells. *The Journal of Physiology*, *594*(17), 4791–4803. https://doi.org/10.1113/jp271667.

Devarasetty, M., Dominijanni, A., Herberg, S., Shelkey, E., Skardal, A., & Soker, S. (2020). Simulating the human colorectal cancer microenvironment in 3D tumor-stroma co-cultures in vitro and in vivo. *Scientific Reports*, *10*(1), 9832. https://doi.org/10.1038/s41598-020-66785-1.

Dolznig, H., Rupp, C., Puri, C., Haslinger, C., Schweifer, N., Wieser, E., et al. (2011). Modeling colon adenocarcinomas in vitro. *The American Journal of Pathology*, *179*(1), 487–501. https://doi.org/10.1016/j.ajpath.2011.03.015.

Domcke, S., Sinha, R., Levine, D. A., Sander, C., & Schultz, N. (2013). Evaluating cell lines as tumour models by comparison of genomic profiles. *Nature Communications*, *4*(1), 2126. https://doi.org/10.1038/ncomms3126.

Drost, J., van Jaarsveld, R. H., Ponsioen, B., Zimberlin, C., van Boxtel, R., Buijs, A., et al. (2015). Sequential cancer mutations in cultured human intestinal stem cells. *Nature*, *521*(7550), 43–47. https://doi.org/10.1038/nature14415.

Drummond, C. G., Bolock, A. M., Ma, C., Luke, C. J., Good, M., & Coyne, C. B. (2017). Enteroviruses infect human enteroids and induce antiviral signaling in a cell lineage-specific manner. *Proceedings of the National Academy of Sciences of the United States of America*, *114*(7), 1672–1677. https://doi.org/10.1073/pnas.1617363114.

Durand, A., Donahue, B., Peignon, G., Letourneur, F., Cagnard, N., Slomianny, C., et al. (2012). Functional intestinal stem cells after Paneth cell ablation induced by the loss of transcription factor Math1 (Atoh1). *Proceedings of the National Academy of Sciences of the United States of America*, *109*(23), 8965–8970. https://doi.org/10.1073/pnas.1201652109.

Ehrbar, M., Rizzi, S. C., Schoenmakers, R. G., Miguel, B. S., Hubbell, J. A., Weber, F. E., et al. (2007). Biomolecular hydrogels formed and degraded via site-specific enzymatic reactions. *Biomacromolecules*, *8*(10), 3000–3007. https://doi.org/10.1021/bm070228f.

Engler, A. J., Sen, S., Sweeney, H. L., & Discher, D. E. (2006). Matrix elasticity directs stem cell lineage specification. *Cell*, *126*(4), 677–689. https://doi.org/10.1016/j.cell.2006.06.044.

Ertel, A., Verghese, A., Byers, S. W., Ochs, M., & Tozeren, A. (2006). Pathway-specific differences between tumor cell lines and normal and tumor tissue cells. *Molecular Cancer*, *5*(1), 55. https://doi.org/10.1186/1476-4598-5-55.

Esch, M. B., Mahler, G. J., Stokol, T., & Shuler, M. L. (2014). Body-on-a-chip simulation with gastrointestinal tract and liver tissues suggests that ingested nanoparticles have the potential to cause liver injury. *Lab on a Chip*, *14*(16), 3081–3092. https://doi.org/10.1039/c4lc00371c.

Evans, G. S., Flint, N., Somers, A. S., Eyden, B., & Potten, C. S. (1992). The development of a method for the preparation of rat intestinal epithelial cell primary cultures. *Journal of Cell Science, 101,* 219–231.

Failli, A., Consolini, R., Legitimo, A., Spisni, R., Castagna, M., Romanini, A., et al. (2009). The challenge of culturing human colorectal tumor cells: Establishment of a cell culture model by the comparison of different methodological approaches. *Tumori, 95*(3), 343–347.

Fairbanks, B. D., Schwartz, M. P., Halevi, A. E., Nuttelman, C. R., Bowman, C. N., & Anseth, K. S. (2009). A versatile synthetic extracellular matrix mimic via Thiol-Norbornene photopolymerization. *Advanced Materials, 21*(48), 5005–5010. https://doi.org/10.1002/adma.200901808.

Fang, Y., & Eglen, R. M. (2017). Three-dimensional cell cultures in drug discovery and development. *SLAS Discovery: Advancing Life Sciences R & D, 22*(5), 456–472. https://doi.org/10.1177/1087057117696795.

Farin, H. F., Van Es, J. H., & Clevers, H. (2012). Redundant sources of Wnt regulate intestinal stem cells and promote formation of Paneth cells. *Gastroenterology, 143*(6), 1518–1529.e1517. https://doi.org/10.1053/j.gastro.2012.08.031.

Fattahi, F., Steinbeck, J. A., Kriks, S., Tchieu, J., Zimmer, B., Kishinevsky, S., et al. (2016). Deriving human ENS lineages for cell therapy and drug discovery in Hirschsprung disease. *Nature, 531*(7592), 105–109. https://doi.org/10.1038/nature16951.

Fearon, E. R., & Vogelstein, B. (1990). A genetic model for colorectal tumorigenesis. *Cell, 61*(5), 759–767. https://doi.org/10.1016/0092-8674(90)90186-I.

Finkbeiner, S. R., Hill, D. R., Altheim, C. H., Dedhia, P. H., Taylor, M. J., Tsai, Y.-H., et al. (2015). Transcriptome-wide analysis reveals hallmarks of human intestine development and maturation in vitro and in vivo. *Stem Cell Reports, 4*(6), 1140–1155. https://doi.org/10.1016/j.stemcr.2015.04.010.

Flanagan, D. J., Pentinmikko, N., Luopajärvi, K., Willis, N. J., Gilroy, K., Raven, A. P., et al. (2021). NOTUM from Apc-mutant cells biases clonal competition to initiate cancer. *Nature.* https://doi.org/10.1038/s41586-021-03525-z.

Frantz, C., Stewart, K. M., & Weaver, V. M. (2010). The extracellular matrix at a glance. *Journal of Cell Science, 123*(Pt. 24), 4195–4200. https://doi.org/10.1242/jcs.023820.

Fujii, M., Matano, M., Toshimitsu, K., Takano, A., Mikami, Y., Nishikori, S., et al. (2018). Human intestinal organoids maintain self-renewal capacity and cellular diversity in Niche-inspired culture condition. *Cell Stem Cell, 23*(6), 787–793.e786. https://doi.org/10.1016/j.stem.2018.11.016.

Fujii, M., Shimokawa, M., Date, S., Takano, A., Matano, M., Nanki, K., et al. (2016). A colorectal tumor organoid library demonstrates progressive loss of niche factor requirements during tumorigenesis. *Cell Stem Cell, 18*(6), 827–838. https://doi.org/10.1016/j.stem.2016.04.003.

Fukamachi, H. (1992). Proliferation and differentiation of fetal rat intestinal epithelial cells in primary serum-free culture. *Journal of Cell Science, 103,* 511–519.

Ganesh, K., Wu, C., O'Rourke, K. P., Szeglin, B. C., Zheng, Y., Sauvé, C. G., et al. (2019). A rectal cancer organoid platform to study individual responses to chemoradiation. *Nature Medicine, 25*(10), 1607–1614. https://doi.org/10.1038/s41591-019-0584-2.

Garabedian, E. M., Roberts, L. J. J., McNevin, M. S., & Gordon, J. I. (1997). Examining the role of paneth cells in the small intestine by lineage ablation in transgenic mice. *Journal of Biological Chemistry, 272*(38), 23729–23740. https://doi.org/10.1074/jbc.272.38.23729.

Gerbe, F., Legraverend, C., & Jay, P. (2012). The intestinal epithelium tuft cells: Specification and function. *Cellular and Molecular Life Sciences: CMLS, 69*(17), 2907–2917. https://doi.org/10.1007/s00018-012-0984-7.

Geurts, M. H., de Poel, E., Amatngalim, G. D., Oka, R., Meijers, F. M., Kruisselbrink, E., et al. (2020). CRISPR-based adenine editors correct nonsense mutations in a cystic fibrosis organoid biobank. *Cell Stem Cell, 26*(4), 503–510.e507. https://doi.org/10.1016/j.stem.2020.01.019.

Gitter, A. H., Wullstein, F., Fromm, M., & Schulzke, J. D. (2001). Epithelial barrier defects in ulcerative colitis: Characterization and quantification by electrophysiological imaging. *Gastroenterology, 121*(6), 1320–1328. https://doi.org/10.1053/gast.2001.29694.

Gribble, F. M., & Reimann, F. (2016). Enteroendocrine cells: Chemosensors in the intestinal epithelium. *Annual Review of Physiology, 78*(1), 277–299. https://doi.org/10.1146/annurev-physiol-021115-105439.

Grivennikov, S. I. (2013). Inflammation and colorectal cancer: Colitis-associated neoplasia. *Seminars in Immunopathology, 35*(2), 229–244. https://doi.org/10.1007/s00281-012-0352-6.

Hall, P. A., Coates, P. J., Ansari, B., & Hopwood, D. (1994). Regulation of cell number in the mammalian gastrointestinal tract: The importance of apoptosis. *Journal of Cell Science, 107*, 3569–3577.

Haramis, A.-P. G., Begthel, H., van den Born, M., van Es, J., Jonkheer, S., Offerhaus, G. J. A., et al. (2004). De novo crypt formation and juvenile polyposis on bmp inhibition in mouse intestine. *Science, 303*(5664), 1684–1686. https://doi.org/10.1126/science.1093587.

Hentze, H., Soong, P. L., Wang, S. T., Phillips, B. W., Putti, T. C., & Dunn, N. R. (2009). Teratoma formation by human embryonic stem cells: Evaluation of essential parameters for future safety studies. *Stem Cell Res, 2*(3), 198–210. https://doi.org/10.1016/j.scr.2009.02.002.

Holly, M. K., & Smith, J. G. (2018). Adenovirus infection of human enteroids reveals interferon sensitivity and preferential infection of goblet cells. *Journal of Virology, 92*(9). https://doi.org/10.1128/jvi.00250-18.

Hou, Q., Ye, L., Liu, H., Huang, L., Yang, Q., Turner, J., et al. (2018). Lactobacillus accelerates ISCs regeneration to protect the integrity of intestinal mucosa through activation of STAT3 signaling pathway induced by LPLs secretion of IL-22. *Cell Death & Differentiation, 25*(9), 1657–1670. https://doi.org/10.1038/s41418-018-0070-2.

Houde, M., Laprise, P., Jean, D., Blais, M., Asselin, C., & Rivard, N. (2001). Intestinal epithelial cell differentiation involves activation of p38 mitogen-activated protein kinase that regulates the homeobox transcription factor CDX2. *Journal of Biological Chemistry, 276*(24), 21885–21894. https://doi.org/10.1074/jbc.M100236200.

Huang, L., Hou, Q., Ye, L., Yang, Q., & Yu, Q. (2017). Crosstalk between H9N2 avian influenza virus and crypt-derived intestinal organoids. *Veterinary Research, 48*(1), 71. https://doi.org/10.1186/s13567-017-0478-6.

Huang, E. H., Hynes, M. J., Zhang, T., Ginestier, C., Dontu, G., Appelman, H., et al. (2009). Aldehyde dehydrogenase 1 is a marker for normal and malignant human colonic stem cells (SC) and tracks SC overpopulation during colon tumorigenesis. *Cancer Research, 69*(8), 3382–3389. https://doi.org/10.1158/0008-5472.Can-08-4418.

Hughes, C. S., Postovit, L. M., & Lajoie, G. A. (2010). Matrigel: A complex protein mixture required for optimal growth of cell culture. *Proteomics, 10*(9), 1886–1890. https://doi.org/10.1002/pmic.200900758.

Imajo, M., Ebisuya, M., & Nishida, E. (2015). Dual role of YAP and TAZ in renewal of the intestinal epithelium. *Nature Cell Biology, 17*(1), 7–19. https://doi.org/10.1038/ncb3084.

Itzkovitz, S., Lyubimova, A., Blat, I. C., Maynard, M., van Es, J., Lees, J., et al. (2011). Single-molecule transcript counting of stem-cell markers in the mouse intestine. *Nature Cell Biology, 14*(1), 106–114. https://doi.org/10.1038/ncb2384.

Jiminez, J. A., Uwiera, T. C., Douglas Inglis, G., & Uwiera, R. R. E. (2015). Animal models to study acute and chronic intestinal inflammation in mammals. *Gut Pathogens, 7*(1), 29. https://doi.org/10.1186/s13099-015-0076-y.

Jung, K. B., Lee, H., Son, Y. S., Lee, M.-O., Kim, Y.-D., Oh, S. J., et al. (2018). Interleukin-2 induces the in vitro maturation of human pluripotent stem cell-derived intestinal organoids. *Nature Communications, 9*(1), 3039. https://doi.org/10.1038/s41467-018-05450-8.

Kaser, A., Zeissig, S., & Blumberg, R. S. (2010). Inflammatory bowel disease. *Annual Review of Immunology, 28*, 573–621. https://doi.org/10.1146/annurev-immunol-030409-101225.

Kasper, S. H., Morell-Perez, C., Wyche, T. P., Sana, T. R., Lieberman, L. A., & Hett, E. C. (2020). Colorectal cancer-associated anaerobic bacteria proliferate in tumor spheroids and alter the microenvironment. *Scientific Reports, 10*(1), 5321. https://doi.org/10.1038/s41598-020-62139-z.

Kemper, K., Prasetyanti, P. R., De Lau, W., Rodermond, H., Clevers, H., & Medema, J. P. (2012). Monoclonal antibodies against Lgr5 identify colorectal cancer stem cells. *Stem Cells, 30*(11), 2378–2386. https://doi.org/10.1002/stem.1233.

Kim, Y.-H., Han, S.-H., Kim, H., Lee, S.-J., Joo, H.-W., Kim, M.-J., et al. (2020). Evaluation of the radiation response and regenerative effects of mesenchymal stem cell-conditioned medium in an intestinal organoid system. *Biotechnology and Bioengineering, 117*(12), 3639–3650. https://doi.org/10.1002/bit.27543.

Kim, Y. S., & Ho, S. B. (2010). Intestinal goblet cells and mucins in health and disease: Recent insights and progress. *Current Gastroenterology Reports, 12*(5), 319–330. https://doi.org/10.1007/s11894-010-0131-2.

Kim, K.-A., Kakitani, M., Zhao, J., Oshima, T., Tang, T., Binnerts, M., et al. (2005). Mitogenic influence of human r-spondin1 on the intestinal epithelium. *Science, 309*(5738), 1256–1259. https://doi.org/10.1126/science.1112521.

Kim, S.-C., Kim, H.-S., Kim, J. H., Jeong, N., Shin, Y.-K., Kim, M. J., et al. (2020). Establishment and characterization of 18 human colorectal cancer cell lines. *Scientific Reports, 10*(1), 6801. https://doi.org/10.1038/s41598-020-63812-z.

Kirov, S., Sasson, A., Zhang, C., Chasalow, S., Dongre, A., Steen, H., et al. (2019). Degradation of the extracellular matrix is part of the pathology of ulcerative colitis. *Molecular Omics, 15*(1), 67–76. https://doi.org/10.1039/C8MO00239H.

Kleinman, H. K., & Martin, G. R. (2005). Matrigel: Basement membrane matrix with biological activity. *Seminars in Cancer Biology, 15*(5), 378–386. https://doi.org/10.1016/j.semcancer.2005.05.004.

Kleinman, H. K., McGarvey, M. L., Hassell, J. R., Star, V. L., Cannon, F. B., Laurie, G. W., et al. (1986). Basement membrane complexes with biological activity. *Biochemistry, 25*(2), 312–318. https://doi.org/10.1021/bi00350a005.

Kolterud, A., Grosse, A. S., Zacharias, W. J., Walton, K. D., Kretovich, K. E., Madison, B. B., et al. (2009). Paracrine Hedgehog signaling in stomach and intestine: New roles for hedgehog in gastrointestinal patterning. *Gastroenterology, 137*(2), 618–628. https://doi.org/10.1053/j.gastro.2009.05.002.

Kondo, J., Endo, H., Okuyama, H., Ishikawa, O., Iishi, H., Tsujii, M., et al. (2011). Retaining cell-cell contact enables preparation and culture of spheroids composed of pure primary cancer cells from colorectal cancer. *Proceedings of the National Academy of Sciences of the United States of America, 108*(15), 6235–6240. https://doi.org/10.1073/pnas.1015938108.

Kong, S. E., Heel, K., McCauley, R., & Hall, J. (1998). The role of enterocytes in gut dysfunction. *Pathology, Research and Practice, 194*(11), 741–751. https://doi.org/10.1016/s0344-0338(98)80063-0.

Kovbasnjuk, O., Zachos, N. C., In, J., Foulke-Abel, J., Ettayebi, K., Hyser, J. M., et al. (2013). Human enteroids: Preclinical models of non-inflammatory diarrhea. *Stem Cell Research & Therapy, 4*(Suppl. 1), S3. https://doi.org/10.1186/scrt364.

Kumagai, T., Rahman, F., & Smith, A. M. (2018). The microbiome and radiation induced-bowel injury: Evidence for potential mechanistic role in disease pathogenesis. *Nutrients*, *10*(10). https://doi.org/10.3390/nu10101405.

Kyd, J. M., & Cripps, A. W. (2008). Functional differences between M cells and enterocytes in sampling luminal antigens. *Vaccine*, *26*(49), 6221–6224. https://doi.org/10.1016/j.vaccine.2008.09.061.

Lahar, N., Lei, N. Y., Wang, J., Jabaji, Z., Tung, S. C., Joshi, V., et al. (2011). Intestinal subepithelial myofibroblasts support in vitro and in vivo growth of human small intestinal epithelium. *PLoS One*, *6*(11). https://doi.org/10.1371/journal.pone.0026898, e26898.

Lamers, M. M., Beumer, J., van der Vaart, J., Knoops, K., Puschhof, J., Breugem, T. I., et al. (2020). SARS-CoV-2 productively infects human gut enterocytes. *Science*, *369*(6499), 50–54. https://doi.org/10.1126/science.abc1669.

Lannagan, T. R. M., Lee, Y. K., Wang, T., Roper, J., Bettington, M. L., Fennell, L., et al. (2019). Genetic editing of colonic organoids provides a molecularly distinct and orthotopic preclinical model of serrated carcinogenesis. *Gut*, *68*(4), 684–692. https://doi.org/10.1136/gutjnl-2017-315920.

Lee, Y. S., Kim, T. Y., Kim, Y., Lee, S. H., Kim, S., Kang, S. W., et al. (2018). Microbiota-derived lactate accelerates intestinal stem-cell-mediated epithelial development. *Cell Host & Microbe*, *24*(6), 833–846.e836. https://doi.org/10.1016/j.chom.2018.11.002.

Lei, N. Y., Jabaji, Z., Wang, J., Joshi, V. S., Brinkley, G. J., Khalil, H., et al. (2014). Intestinal subepithelial myofibroblasts support the growth of intestinal epithelial stem cells. *PLoS One*, *9*(1), e84651. https://doi.org/10.1371/journal.pone.0084651.

Leibowitz, B. J., Yang, L., Wei, L., Buchanan, M. E., Rachid, M., Parise, R. A., et al. (2018). Targeting p53-dependent stem cell loss for intestinal chemoprotection. *Science Translational Medicine*, *10*(427). https://doi.org/10.1126/scitranslmed.aam7610.

Li, X., Sun, Q., Li, Q., Kawazoe, N., & Chen, G. (2018). Functional hydrogels with tunable structures and properties for tissue engineering applications. *Frontiers in Chemistry*, *6*, 499. https://doi.org/10.3389/fchem.2018.00499.

Liang, L., Shen, L., Fu, G., Yao, Y., Li, G., Deng, Y., et al. (2020). Regulation of the regeneration of intestinal stem cells after irradiation. *Annals of Translational Medicine*, *8*(17), 1063. https://doi.org/10.21037/atm-20-4542.

Lichtenstein, G. R., & Rutgeerts, P. (2010). Importance of mucosal healing in ulcerative colitis. *Inflammatory Bowel Diseases*, *16*(2), 338–346. https://doi.org/10.1002/ibd.20997.

Lindemans, C. A., Calafiore, M., Mertelsmann, A. M., O'Connor, M. H., Dudakov, J. A., Jenq, R. R., et al. (2015). Interleukin-22 promotes intestinal-stem-cell-mediated epithelial regeneration. *Nature*, *528*(7583), 560–564. https://doi.org/10.1038/nature16460.

Liu, R., Moriggl, R., Zhang, D., Li, H., Karns, R., Ruan, H.-B., et al. (2019). Constitutive STAT5 activation regulates Paneth and Paneth-like cells to control *Clostridium difficile* colitis. *Life Science Alliance*, *2*(2), e201900296. https://doi.org/10.26508/lsa.201900296.

Lu, L., Li, W., Chen, L., Su, Q., Wang, Y., Guo, Z., et al. (2019). Radiation-induced intestinal damage: Latest molecular and clinical developments. *Future Oncology*, *15*(35), 4105–4118. https://doi.org/10.2217/fon-2019-0416.

Martinez Rodriguez, N. R., Eloi, M. D., Huynh, A., Dominguez, T., Lam, A. H., Carcamo-Molina, D., et al. (2012). Expansion of Paneth cell population in response to enteric Salmonella enterica serovar Typhimurium infection. *Infection and Immunity*, *80*(1), 266–275. https://doi.org/10.1128/iai.05638-11.

Matano, M., Date, S., Shimokawa, M., Takano, A., Fujii, M., Ohta, Y., et al. (2015). Modeling colorectal cancer using CRISPR-Cas9–mediated engineering of human intestinal organoids. *Nature Medicine*, *21*(3), 256–262. https://doi.org/10.1038/nm.3802.

McCracken, K. W., Howell, J. C., Wells, J. M., & Spence, J. R. (2011). Generating human intestinal tissue from pluripotent stem cells in vitro. *Nature Protocols*, *6*(12), 1920–1928. https://doi.org/10.1038/nprot.2011.410.

McElroy, S. J., Frey, M. R., Torres, B. A., & Maheshwari, A. (2018). 72—Innate and mucosal immunity in the developing gastrointestinal tract. In C. A. Gleason, & S. E. Juul (Eds.), *Avery's diseases of the newborn* (10th ed., pp. 1054–1067.e1055). Philadelphia: Elsevier.

Middendorp, S., Schneeberger, K., Wiegerinck, C. L., Mokry, M., Akkerman, R. D. L., van Wijngaarden, S., et al. (2014). Adult stem cells in the small intestine are intrinsically programmed with their location-specific function: Intestinal stem cells are intrinsically programmed. *Stem Cells*, *32*(5), 1083–1091. https://doi.org/10.1002/stem.1655.

Mithal, A., Capilla, A., Heinze, D., Berical, A., Villacorta-Martin, C., Vedaie, M., et al. (2020). Generation of mesenchyme free intestinal organoids from human induced pluripotent stem cells. *Nature Communications*, *11*(1), 215. https://doi.org/10.1038/s41467-019-13916-6.

Miyazono, K., Kamiya, Y., & Morikawa, M. (2010). Bone morphogenetic protein receptors and signal transduction. *Journal of Biochemistry*, *147*(1), 35–51. https://doi.org/10.1093/jb/mvp148.

Mori-Akiyama, Y., van den Born, M., van Es, J. H., Hamilton, S. R., Adams, H. P., Zhang, J., et al. (2007). SOX9 is required for the differentiation of paneth cells in the intestinal epithelium. *Gastroenterology*, *133*(2), 539–546. https://doi.org/10.1053/j.gastro.2007.05.020.

Moser, A. R., Pitot, H. C., & Dove, W. F. (1990). A dominant mutation that predisposes to multiple intestinal neoplasia in the mouse. *Science*, *247*(4940), 322–324.

Mouradov, D., Sloggett, C., Jorissen, R. N., Love, C. G., Li, S., Burgess, A. W., et al. (2014). Colorectal cancer cell lines are representative models of the main molecular subtypes of primary cancer. *Cancer Research*, *74*(12), 3238–3247. https://doi.org/10.1158/0008-5472.CAN-14-0013.

Muñoz, J., Stange, D. E., Schepers, A. G., van de Wetering, M., Koo, B.-K., Itzkovitz, S., et al. (2012). The Lgr5 intestinal stem cell signature: Robust expression of proposed quiescent '+4' cell markers. *The EMBO Journal*, *31*(14), 3079–3091. https://doi.org/10.1038/emboj.2012.166.

Nag, D., Bhanja, P., Riha, R., Sanchez-Guerrero, G., Kimler, B. F., Tsue, T. T., et al. (2019). Auranofin protects intestine against radiation injury by modulating p53/p21 pathway and radiosensitizes human colon tumor. *Clinical Cancer Research: An Official Journal of the American Association for Cancer Research*, *25*(15), 4791–4807. https://doi.org/10.1158/1078-0432.ccr-18-2751.

Nguyen, K. T., & West, J. L. (2002). Photopolymerizable hydrogels for tissue engineering applications. *Biomaterials*, *23*(22), 4307–4314. https://doi.org/10.1016/s0142-9612(02)00175-8.

Nielsen, O. H., & Ainsworth, M. A. (2013). Tumor necrosis factor inhibitors for inflammatory bowel disease. *The New England Journal of Medicine*, *369*(8), 754–762. https://doi.org/10.1056/NEJMct1209614.

Noel, G., Baetz, N. W., Staab, J. F., Donowitz, M., Kovbasnjuk, O., Pasetti, M. F., et al. (2017). A primary human macrophage-enteroid co-culture model to investigate mucosal gut physiology and host-pathogen interactions. *Scientific Reports*, *7*(1), 45270. https://doi.org/10.1038/srep45270.

Nozaki, K., Mochizuki, W., Matsumoto, Y., Matsumoto, T., Fukuda, M., Mizutani, T., et al. (2016). Co-culture with intestinal epithelial organoids allows efficient expansion and motility analysis of intraepithelial lymphocytes. *Journal of Gastroenterology*, *51*(3), 206–213. https://doi.org/10.1007/s00535-016-1170-8.

Nyga, A., Loizidou, M., Emberton, M., & Cheema, U. (2013). A novel tissue engineered three-dimensional in vitro colorectal cancer model. *Acta Biomaterialia, 9*(8), 7917–7926. https://doi.org/10.1016/j.actbio.2013.04.028.

O'Brien, C. A., Pollett, A., Gallinger, S., & Dick, J. E. (2007). A human colon cancer cell capable of initiating tumour growth in immunodeficient mice. *Nature, 445*(7123), 106–110. https://doi.org/10.1038/nature05372.

Ootani, A., Li, X., Sangiorgi, E., Ho, Q. T., Ueno, H., Toda, S., et al. (2009). Sustained in vitro intestinal epithelial culture within a Wnt-dependent stem cell niche. *Nature Medicine, 15*(6), 701–706. https://doi.org/10.1038/nm.1951.

Orkin, R. W., Gehron, P., McGoodwin, E. B., Martin, G. R., Valentine, T., & Swarm, R. (1977). A murine tumor producing a matrix of basement membrane. *Journal of Experimental Medicine, 145*(1), 204–220. https://doi.org/10.1084/jem.145.1.204.

Otsuka, M., Kang, Y. J., Ren, J., Jiang, H., Wang, Y., Omata, M., et al. (2010). Distinct effects of p38α deletion in myeloid lineage and gut epithelia in mouse models of inflammatory bowel disease. *Gastroenterology, 138*(4), 1255–1265.e1259. https://doi.org/10.1053/j.gastro.2010.01.005.

Pang, R., Law, W. L., Chu, A. C., Poon, J. T., Lam, C. S., Chow, A. K., et al. (2010). A subpopulation of CD26+ cancer stem cells with metastatic capacity in human colorectal cancer. *Cell Stem Cell, 6*(6), 603–615. https://doi.org/10.1016/j.stem.2010.04.001.

Park, C. S., Nguyen, L. P., & Yong, D. (2020). Development of colonic organoids containing enteric nerves or blood vessels from human embryonic stem cells. *Cells, 9*(10), 2209. Retrieved from https://www.mdpi.com/2073-4409/9/10/2209.

Pastuła, A., Middelhoff, M., Brandtner, A., Tobiasch, M., Höhl, B., Nuber, A. H., et al. (2016). Three-dimensional gastrointestinal organoid culture in combination with nerves or fibroblasts: A method to characterize the gastrointestinal stem cell niche. *Stem Cells International, 2016*, 1–16. https://doi.org/10.1155/2016/3710836.

Pinchuk, I. V., Mifflin, R. C., Saada, J. I., & Powell, D. W. (2010). Intestinal mesenchymal cells. *Current Gastroenterology Reports, 12*(5), 310–318. https://doi.org/10.1007/s11894-010-0135-y.

Pinto, D., Gregorieff, A., Begthel, H., & Clevers, H. (2003). Canonical Wnt signals are essential for homeostasis of the intestinal epithelium. *Genes & Development, 17*(14), 1709–1713. https://doi.org/10.1101/gad.267103.

Potten, C. S. (1977). Extreme sensitivity of some intestinal crypt cells to X and γ irradiation. *Nature, 269*(5628), 518–521. https://doi.org/10.1038/269518a0.

Potten, C. S., Booth, C., Tudor, G. L., Booth, D., Brady, G., Hurley, P., et al. (2003). Identification of a putative intestinal stem cell and early lineage marker; musashi-1. *Differentiation, 71*(1), 28–41. https://doi.org/10.1046/j.1432-0436.2003.700603.x.

Powell, D. W., Pinchuk, I. V., Saada, J. I., Chen, X., & Mifflin, R. C. (2011). Mesenchymal cells of the intestinal lamina propria. *Annual Review of Physiology, 73*, 213–237. https://doi.org/10.1146/annurev.physiol.70.113006.100646.

Powell, A. E., Wang, Y., Li, Y., Poulin, E. J., Means, A. L., Washington, M. K., et al. (2012). The pan-ErbB negative regulator Lrig1 is an intestinal stem cell marker that functions as a tumor suppressor. *Cell, 149*(1), 146–158. https://doi.org/10.1016/j.cell.2012.02.042.

Punkenburg, E., Vogler, T., Büttner, M., Amann, K., Waldner, M., Atreya, R., et al. (2016). Batf-dependent Th17 cells critically regulate IL-23 driven colitis-associated colon cancer. *Gut, 65*(7), 1139. https://doi.org/10.1136/gutjnl-2014-308227.

Qi, Z., Li, Y., Zhao, B., Xu, C., Liu, Y., Li, H., et al. (2017). BMP restricts stemness of intestinal Lgr5+ stem cells by directly suppressing their signature genes. *Nature Communications, 8*(1), 13824. https://doi.org/10.1038/ncomms13824.

Rahmani, S., Breyner, N. M., Su, H.-M., Verdu, E. F., & Didar, T. F. (2019). Intestinal organoids: A new paradigm for engineering intestinal epithelium in vitro. *Biomaterials, 194*, 195–214. https://doi.org/10.1016/j.biomaterials.2018.12.006.

Rawla, P., Sunkara, T., & Barsouk, A. (2019). Epidemiology of colorectal cancer: Incidence, mortality, survival, and risk factors. *Gastroenterology Review, 14*(2), 89–103. https://doi.org/10.5114/pg.2018.81072.

Reddy, V. K., Short, S. P., Barrett, C. W., Mittal, M. K., Keating, C. E., Thompson, J. J., et al. (2016). BVES regulates intestinal stem cell programs and intestinal crypt viability after radiation. *Stem Cells, 34*(6), 1626–1636. https://doi.org/10.1002/stem.2307.

Ricci-Vitiani, L., Lombardi, D. G., Pilozzi, E., Biffoni, M., Todaro, M., Peschle, C., et al. (2007). Identification and expansion of human colon-cancer-initiating cells. *Nature, 445*(7123), 111–115. https://doi.org/10.1038/nature05384.

Sandborn, W. J., & Hanauer, S. B. (1999). Antitumor necrosis factor therapy for inflammatory bowel disease: A review of agents, pharmacology, clinical results, and safety. *Inflammatory Bowel Diseases, 5*(2), 119–133. https://doi.org/10.1097/00054725-199905000-00008.

Sangiorgi, E., & Capecchi, M. R. (2008). Bmi1 is expressed in vivo in intestinal stem cells. *Nature Genetics, 40*(7), 915–920. https://doi.org/10.1038/ng.165.

Sato, T., Stange, D. E., Ferrante, M., Vries, R. G. J., van Es, J. H., van den Brink, S., et al. (2011). Long-term expansion of epithelial organoids from human colon, adenoma, adenocarcinoma, and barrett's epithelium. *Gastroenterology, 141*(5), 1762–1772. https://doi.org/10.1053/j.gastro.2011.07.050.

Sato, T., van Es, J. H., Snippert, H. J., Stange, D. E., Vries, R. G., van den Born, M., et al. (2011). Paneth cells constitute the niche for Lgr5 stem cells in intestinal crypts. *Nature, 469*(7330), 415–418. https://doi.org/10.1038/nature09637.

Sato, T., Vries, R. G., Snippert, H. J., van de Wetering, M., Barker, N., Stange, D. E., et al. (2009). Single Lgr5 stem cells build crypt-villus structures in vitro without a mesenchymal niche. *Nature, 459*(7244), 262–265. https://doi.org/10.1038/nature07935.

Saxena, K., Blutt, S. E., Ettayebi, K., Zeng, X. L., Broughman, J. R., Crawford, S. E., et al. (2016). Human intestinal enteroids: A new model to study human rotavirus infection, host restriction, and pathophysiology. *Journal of Virology, 90*(1), 43–56. https://doi.org/10.1128/jvi.01930-15.

Schepers, A. G., Snippert, H. J., Stange, D. E., van den Born, M., van Es, J. H., van de Wetering, M., et al. (2012). Lineage tracing reveals Lgr5 + stem cell activity in mouse intestinal adenomas. *Science, 337*(6095), 730–735. https://doi.org/10.1126/science.1224676.

Schmitz, H., Fromm, M., Bentzel, C. J., Scholz, P., Detjen, K., Mankertz, J., et al. (1999). Tumor necrosis factor-alpha (TNFalpha) regulates the epithelial barrier in the human intestinal cell line HT-29/B6. *Journal of Cell Science, 112*(Pt. 1), 137–146.

Schwank, G., Koo, B.-K., Sasselli, V., Dekkers, J. F., Heo, I., Demircan, T., et al. (2013). Functional repair of CFTR by CRISPR/Cas9 in intestinal stem cell organoids of cystic fibrosis patients. *Cell Stem Cell, 13*(6), 653–658. https://doi.org/10.1016/j.stem.2013.11.002.

Shadad, A. K., Sullivan, F. J., Martin, J. D., & Egan, L. J. (2013). Gastrointestinal radiation injury: Prevention and treatment. *World Journal of Gastroenterology, 19*(2), 199–208. https://doi.org/10.3748/wjg.v19.i2.199.

Shimokawa, M., Ohta, Y., Nishikori, S., Matano, M., Takano, A., Fujii, M., et al. (2017). Visualization and targeting of LGR5(+) human colon cancer stem cells. *Nature, 545*(7653), 187–192. https://doi.org/10.1038/nature22081.

Sicard, J.-F., Le Bihan, G., Vogeleer, P., Jacques, M., & Harel, J. (2017). Interactions of intestinal bacteria with components of the intestinal mucus. *Frontiers in Cellular and Infection Microbiology, 7*, 387. https://doi.org/10.3389/fcimb.2017.00387.

Snippert, H. J., van der Flier, L. G., Sato, T., van Es, J. H., van den Born, M., Kroon-Veenboer, C., et al. (2010). Intestinal crypt homeostasis results from neutral competition between symmetrically dividing Lgr5 stem cells. *Cell, 143*(1), 134–144. https://doi.org/10.1016/j.cell.2010.09.016.

Snoeck, V., Goddeeris, B., & Cox, E. (2005). The role of enterocytes in the intestinal barrier function and antigen uptake. *Microbes and Infection, 7*(7-8), 997–1004. https://doi.org/10. 1016/j.micinf.2005.04.003.

Soroosh, A., Albeiroti, S., West, G. A., Willard, B., Fiocchi, C., & de la Motte, C. A. (2016). Crohn's disease fibroblasts overproduce the novel protein KIAA1199 to create proinflammatory hyaluronan fragments. *Cellular and Molecular Gastroenterology and Hepatology, 2*(3), 358–368.e354. https://doi.org/10.1016/j.jcmgh.2015.12.007.

Spence, J. R., Mayhew, C. N., Rankin, S. A., Kuhar, M. F., Vallance, J. E., Tolle, K., et al. (2011). Directed differentiation of human pluripotent stem cells into intestinal tissue in vitro. *Nature, 470*(7332), 105–109. https://doi.org/10.1038/nature09691.

Spicer, C. D. (2020). Hydrogel scaffolds for tissue engineering: The importance of polymer choice. *Polymer Chemistry, 11*(2), 184–219. https://doi.org/10.1039/C9PY01021A.

Sugimoto, S., Ohta, Y., Fujii, M., Matano, M., Shimokawa, M., Nanki, K., et al. (2018). Reconstruction of the human colon epithelium in vivo. *Cell Stem Cell, 22*(2), 171–176.e175. https://doi.org/10.1016/j.stem.2017.11.012.

Takeda, N., Jain, R., LeBoeuf, M. R., Wang, Q., Lu, M. M., & Epstein, J. A. (2011). Interconversion between intestinal stem cell populations in distinct niches. *Science, 334*(6061), 1420–1424. https://doi.org/10.1126/science.1213214.

Talbot, N. C., & Caperna, T. J. (2015). Proteome array identification of bioactive soluble proteins/peptides in Matrigel: Relevance to stem cell responses. *Cytotechnology, 67*(5), 873–883. https://doi.org/10.1007/s10616-014-9727-y.

Tao, G., Huang, J., Moorthy, B., Wang, C., Hu, M., Gao, S., et al. (2020). Potential role of drug metabolizing enzymes in chemotherapy-induced gastrointestinal toxicity and hepatotoxicity. *Expert Opinion on Drug Metabolism & Toxicology, 16*(11), 1109–1124. https://doi.org/10.1080/17425255.2020.1815705.

Tentler, J. J., Tan, A. C., Weekes, C. D., Jimeno, A., Leong, S., Pitts, T. M., et al. (2012). Patient-derived tumour xenografts as models for oncology drug development. *Nature Reviews. Clinical Oncology, 9*(6), 338–350. https://doi.org/10.1038/nrclinonc. 2012.61.

Tirado, F. R., Bhanja, P., Castro-Nallar, E., Olea, X. D., Salamanca, C., & Saha, S. (2021). Radiation-induced toxicity in rectal epithelial stem cell contributes to acute radiation injury in rectum. *Stem Cell Research & Therapy, 12*(1), 63. https://doi.org/10.1186/ s13287-020-02111-w.

Todaro, M., Alea, M. P., Di Stefano, A. B., Cammareri, P., Vermeulen, L., Iovino, F., et al. (2007). Colon cancer stem cells dictate tumor growth and resist cell death by production of interleukin-4. *Cell Stem Cell, 1*(4), 389–402. https://doi.org/10.1016/j.stem. 2007.08.001.

Todaro, M., Gaggianesi, M., Catalano, V., Benfante, A., Iovino, F., Biffoni, M., et al. (2014). CD44v6 is a marker of constitutive and reprogrammed cancer stem cells driving colon cancer metastasis. *Cell Stem Cell, 14*(3), 342–356. https://doi.org/10.1016/j.stem.2014. 01.009.

Tsai, Y.-H., Nattiv, R., Dedhia, P. H., Nagy, M. S., Chin, A. M., Thomson, M., et al. (2017). In vitro patterning of pluripotent stem cell-derived intestine recapitulates in vivo human development. *Development (Cambridge, England), 144*(6), 1045–1055. https://doi.org/10.1242/dev.138453.

van de Wetering, M., Francies, H. E., Francis, J. M., Bounova, G., Iorio, F., Pronk, A., et al. (2015). Prospective derivation of a living organoid biobank of colorectal cancer patients. *Cell, 161*(4), 933–945. https://doi.org/10.1016/j.cell.2015.03.053.

van der Flier, L. G., & Clevers, H. (2009). Stem cells, self-renewal, and differentiation in the intestinal epithelium. *Annual Review of Physiology, 71*(1), 241–260. https://doi.org/ 10.1146/annurev.physiol.010908.163145.

van der Flier, L. G., Haegebarth, A., Stange, D. E., van de Wetering, M., & Clevers, H. (2009). OLFM4 is a robust marker for stem cells in human intestine and marks a subset of colorectal cancer cells. *Gastroenterology, 137*(1), 15–17. https://doi.org/10.1053/j.gastro.2009.05.035.

van der Flier, L. G., van Gijn, M. E., Hatzis, P., Kujala, P., Haegebarth, A., Stange, D. E., et al. (2009). Transcription factor achaete scute-like 2 controls intestinal stem cell fate. *Cell, 136*(5), 903–912. https://doi.org/10.1016/j.cell.2009.01.031.

van Es, J. H., Wiebrands, K., López-Iglesias, C., van de Wetering, M., Zeinstra, L., van den Born, M., et al. (2019). Enteroendocrine and tuft cells support Lgr5 stem cells on Paneth cell depletion. *Proceedings of the National Academy of Sciences of the United States of America, 116*(52), 26599–26605. https://doi.org/10.1073/pnas.1801888117.

Van Landeghem, L., Santoro, M. A., Krebs, A. E., Mah, A. T., Dehmer, J. J., Gracz, A. D., et al. (2012). Activation of two distinct Sox9-EGFP-expressing intestinal stem cell populations during crypt regeneration after irradiation. *The American Journal of Physiology-Gastrointestinal and Liver Physiology, 302*(10), G1111–G1132. https://doi.org/10.1152/ajpgi.00519.2011.

van Neerven, S. M., de Groot, N. E., Nijman, L. E., Scicluna, B. P., van Driel, M. S., Lecca, M. C., et al. (2021). Apc-mutant cells act as supercompetitors in intestinal tumour initiation. *Nature.* https://doi.org/10.1038/s41586-021-03558-4.

Velasco, V., Shariati, S. A., & Esfandyarpour, R. (2020). Microtechnology-based methods for organoid models. *Microsystems & Nanoengineering, 6*(1), 76. https://doi.org/10.1038/s41378-020-00185-3.

Verissimo, C. S., Overmeer, R. M., Ponsioen, B., Drost, J., Mertens, S., Verlaan-Klink, I., et al. (2016). Targeting mutant RAS in patient-derived colorectal cancer organoids by combinatorial drug screening. *eLife, 5*, e18489. https://doi.org/10.7554/eLife.18489.

Watson, C. L., Mahe, M. M., Múnera, J., Howell, J. C., Sundaram, N., Poling, H. M., et al. (2014). An in vivo model of human small intestine using pluripotent stem cells. *Nature Medicine, 20*(11), 1310–1314. https://doi.org/10.1038/nm.3737.

Whitehead, R. H., Demmler, K., Rockman, S. P., & Watson, N. K. (1999). Clonogenic growth of epithelial cells from normal colonic mucosa from both mice and humans. *Gastroenterology, 117*(4), 858–865. https://doi.org/10.1016/S0016-5085(99)70344-6.

Workman, M. J., Mahe, M. M., Trisno, S., Poling, H. M., Watson, C. L., Sundaram, N., et al. (2017). Engineered human pluripotent-stem-cell-derived intestinal tissues with a functional enteric nervous system. *Nature Medicine, 23*(1), 49–59. https://doi.org/10.1038/nm.4233.

Yamauchi, M., Otsuka, K., Kondo, H., Hamada, N., Tomita, M., Takahashi, M., et al. (2014). A novel in vitro survival assay of small intestinal stem cells after exposure to ionizing radiation. *Journal of Radiation Research, 55*(2), 381–390. https://doi.org/10.1093/jrr/rrt123.

Yan, K. S., Chia, L. A., Li, X., Ootani, A., Su, J., Lee, J. Y., et al. (2012). The intestinal stem cell markers Bmi1 and Lgr5 identify two functionally distinct populations. *Proceedings of the National Academy of Sciences of the United States of America, 109*(2), 466–471. https://doi.org/10.1073/pnas.1118857109.

Yilmaz, Ö. H., Katajisto, P., Lamming, D. W., Gültekin, Y., Bauer-Rowe, K. E., Sengupta, S., et al. (2012). mTORC1 in the Paneth cell niche couples intestinal stem-cell function to calorie intake. *Nature, 486*(7404), 490–495. https://doi.org/10.1038/nature11163.

Yu, J. (2013). Intestinal stem cell injury and protection during cancer therapy. *Translational Cancer Research, 2*(5), 384–396. Retrieved from https://pubmed.ncbi.nlm.nih.gov/24683536.

Yu, S., Tong, K., Zhao, Y., Balasubramanian, I., Yap, G. S., Ferraris, R. P., et al. (2018). Paneth cell multipotency induced by notch activation following injury. *Cell Stem Cell*, *23*(1), 46–59.e45. https://doi.org/10.1016/j.stem.2018.05.002.

Yui, S., Azzolin, L., Maimets, M., Pedersen, M. T., Fordham, R. P., Hansen, S. L., et al. (2018). YAP/TAZ-dependent reprogramming of colonic epithelium links ECM remodeling to tissue regeneration. *Cell Stem Cell*, *22*(1), 35–49.e37. https://doi.org/10. 1016/j.stem.2017.11.001.

Yui, S., Nakamura, T., Sato, T., Nemoto, Y., Mizutani, T., Zheng, X., et al. (2012). Functional engraftment of colon epithelium expanded in vitro from a single adult Lgr5$^+$ stem cell. *Nature Medicine*, *18*(4), 618–623. https://doi.org/10.1038/nm.2695.

Yum, M. K., Han, S., Fink, J., Wu, S.-H. S., Dabrowska, C., Trendafilova, T., et al. (2021). Tracing oncogene-driven remodelling of the intestinal stem cell niche. *Nature*. https:// doi.org/10.1038/s41586-021-03605-0.

Zaborin, A., Krezalek, M., Hyoju, S., Defazio, J. R., Setia, N., Belogortseva, N., et al. (2017). Critical role of microbiota within cecal crypts on the regenerative capacity of the intestinal epithelium following surgical stress. *The American Journal of Physiology-Gastrointestinal and Liver Physiology*, *312*(2), G112–g122. https://doi.org/10. 1152/ajpgi.00294.2016.

Zaborin, A., Penalver Bernabe, B., Keskey, R., Sangwan, N., Hyoju, S., Gottel, N., et al. (2020). Spatial compartmentalization of the microbiome between the lumen and crypts is lost in the murine cecum following the process of surgery, including overnight fasting and exposure to antibiotics. *mSystems*, *5*(3). https://doi.org/10.1128/ mSystems.00377-20. e00377-00320.

Zacharias, W. J., Madison, B. B., Kretovich, K. E., Walton, K. D., Richards, N., Udager, A. M., et al. (2011). Hedgehog signaling controls homeostasis of adult intestinal smooth muscle. *Developmental Biology*, *355*(1), 152–162. https://doi.org/10.1016/ j.ydbio.2011.04.025.

Zachos, N. C., Kovbasnjuk, O., Foulke-Abel, J., In, J., Blutt, S. E., de Jonge, H. R., et al. (2016). Human enteroids/colonoids and intestinal organoids functionally recapitulate normal intestinal physiology and pathophysiology. *Journal of Biological Chemistry*, *291*(8), 3759–3766. https://doi.org/10.1074/jbc.R114.635995.

Zeissig, S., Bojarski, C., Buergel, N., Mankertz, J., Zeitz, M., Fromm, M., et al. (2004). Downregulation of epithelial apoptosis and barrier repair in active Crohn's disease by tumour necrosis factor alpha antibody treatment. *Gut*, *53*(9), 1295–1302. https://doi. org/10.1136/gut.2003.036632.

Zhang, X. T., Gong, A. Y., Wang, Y., Chen, X., Lim, S. S., Dolata, C. E., et al. (2016). Cryptosporidium parvum infection attenuates the ex vivo propagation of murine intestinal enteroids. *Physiological Reports*, *4*(24). https://doi.org/10.14814/phy2.13060.

CHAPTER THREE

Reconstructing the lung stem cell niche *in vitro*

Dayanand Swami[a,†], **Jyotirmoi Aich**[a,†], **Bharti Bisht**[b],
and Manash K. Paul[c,*]

[a]School of Biotechnology and Bioinformatics, D. Y. Patil Deemed to Be University, CBD Belapur, Navi Mumbai, Maharashtra, India
[b]Division of Thoracic Surgery, David Geffen School of Medicine, University of California Los Angeles, Los Angeles, CA, United States
[c]Division of Pulmonary and Critical Care Medicine, David Geffen School of Medicine, University of California Los Angeles (UCLA), Los Angeles, CA, United States
*Corresponding author: e-mail addresses: paul_cancerbiotech@yahoo.co.in; manashp@ucla.edu

Contents

1. Introduction	99
2. Lung development	100
3. Lung stem cell diversity	102
4. Lung regeneration	107
5. Lung stem cell niche	109
5.1 Early development of the lung and niche	110
5.2 Proximal airway stem cell niche in adult lung	110
5.3 Distal stem cell niche in adult lung	111
5.4 Niche and lung diseases	115
6. Recapitulating stem cell niche	116
6.1 ECM topography and material	118
6.2 Scaffold biomaterial, natural and synthetic	119
7. Culture systems and morphogens to study lung stem cell niche	120
7.1 2D and 3D cell culture system	120
7.2 Organoids	121
7.3 Morphogens and peptide	122
8. Artificial scaffold and lung stem/progenitor cell niche	124
8.1 Scaffold-based engineered lung organoid	124
8.2 Bioprinting	127
9. Conclusion and future direction	131
Acknowledgments	132
Competing interests	132
References	133

---

† Equal contribution.

Advances in Stem Cells and their Niches, Volume 6
ISSN 2468-5097
https://doi.org/10.1016/bs.asn.2022.05.001

97

Abbreviations

3D	three-dimensional
ABSC	airway basal stem cell
ASC	adult stem cell
ASMC	airway smooth muscle cell
AT1	alveolar type I cell
AT2	alveolar type II cell
BADJ	bronchoalveolar duct junction
BASC	bronchoalveolar stem cells
BMDM	bone marrow-derived macrophage
BPD	bronchopulmonary dysplasia
COPD	chronic obstructive pulmonary disease
CPP	cell-penetrating peptide
DASC	distal alveolar stem cell
DATP	damage-associated transient progenitors
DIPNECH	diffuse idiopathic pulmonary neuroendocrine cell hyperplasia
ECM	extracellular matrix
EPC	endothelial progenitor cell
ESC	embryonic stem cell
FN	fibronectin
FRESH	freeform reversible embedding of suspended hydrogels
GAG	glycosaminoglycan
hPSC	human pluripotent stem cell
HS	heparin sulfate
ICZ	inter-cartligeneous zone
iPSC	induced pluripotent stem cell
LGDW	laser-guided direct writing
LIF	lipofibroblast
LNEP	lineage negative epithelial precursor
MSC	mesenchymal stem cell
NE	neuroendocrine
NEB	neuroepithelial bodies
NEHI	neuroendocrine hyperplasia of infancy
PEG	polyethylene glycol
PG	proteoglycan
PLG	poly(lactide-*co*-glycolide)
PNEC	pulmonary neuroendocrine cell
PNX	pneumonectomy
scRNAseq	single-cell RNA-sequencing
SLA	stereolithography
SMG	submucosal gland

1. Introduction

The lungs are a vital part of our respiratory system and carry out the vital physiological function of respiration, a process of gaseous exchange and blood oxygenation. The right and the left lungs are separated into smaller units called lobes. The human right lung comprises three lobes known as the superior, middle, and inferior lobes, while the left lung is composed of just two lobes, superior and inferior lobes. The lobes are further separated into bronchopulmonary segments and comprise distinct anatomic zones. The bronchi split further, eventually forming bronchioles <1 mm in diameter. Each bronchiole is divided into 50–80 terminal bronchioles, which are the respiratory bronchiolar branches, respiratory bronchioles, alveolar ducts, alveolar sacs, and alveoli are all included in the acinus, which is the terminal respiratory unit. The respiratory system's structural and functional components are the alveoli. An adult person has roughly 300 million alveoli, which equates to around 80 square meters of surface area for gaseous exchange. Lung disease is a severe public health concern in the United States, with over 400,000 people dying from it yearly (Labaki & Han, 2020). Obstructive or restrictive lung diseases like COPD cause conditions that make it hard to completely exhale the air within the lungs. Pulmonary fibrosis, sarcoidosis, and pulmonary hypertension have a prominent histopathological feature of malformation or loss of the alveolar gas-exchange areas. Although each condition has various therapeutic options, lung transplantation is often preferred as the only permanent solution. Lung transplantation has a few drawbacks like organ rejection, low donor organ availability, chance of post-transplant infection, and demonstrates poor prognosis in patients (Shah, Rahulan, Kumar, Dutta, & Attawar, 2021; Van Raemdonck et al., 2009).

Even though most organs in the adult human body can maintain themselves and undergo repair after injury, the lung is a very quiescent tissue, long considered to have a low reparative ability and sensitivity to scarring. It is well established now that the lung has a remarkable ability to repair itself, and these processes are primarily reliant on stem cells. Many tissue-specific stem/progenitor cells may be found in distinct anatomic areas of normal lungs and undergo self-renewal, division, and differentiation to replace old or injured lung cells and maintain lung homeostasis. Impairment of regenerative ability of the stem/progenitor cells may occur due to lung injury, disease, and age-related impairment. Lung tissues have facultative stem/progenitor cell populations with regenerative potential that can be activated to maintain tissue homeostasis following tissue damage. This response is distinct from organs such as the gut and hematological system, which have rapid cellular turnover and need a specialized and well-defined undifferentiated stem cell population, or organs like the heart and brain, which have minimal ability to regenerate even after damage. Classically stem cells are defined as cells that can self-renew indefinitely and generate clonal populations of cells. The mechanism of lung maintenance, repair, and regeneration may appear similar to other endodermally-derived epithelial tissue/organs such as the pancreas or the liver (Kotton & Morrisey, 2014).

Considering the regenerative potentials of lung stem/progenitor cells, induced stem cell pluripotency (iPSCs), directed differentiation, stem cell therapies, implanted tissue constructions, and tissue engineering will be valuable in providing life-saving solutions to these challenging clinical situations. For the sake of simplicity, regenerative medicine or tissue-engineering technique may be thought of as any combination of four essential elements: scaffolds, cells, and matrices, growth factors or other signaling molecules, and three-dimensional cell culture systems grown in bioreactors (Han et al., 2020; Stephenson & Grayson, 2018). In this chapter, we will discuss lung stem cell biology basics before delving into all four of these essential concepts and recapitulating the lung stem cell niche *in vitro*.

Over the last 10 years, cell therapies and bioengineering techniques for lung disorders have advanced at a breakneck pace. Study and use of mesenchymal stem cells (MSCs) (stromal) and endothelial progenitor cells (EPCs) having immunomodulatory and paracrine effects, as well as the rapidly developing area of *ex vivo* lung bioengineering (De Santis, Bölükbas, Lindstedt, & Wagner, 2018; Whitsett & Weaver, 2002; Wilkinson et al., 2017), have largely replaced the initial focus on structural engraftment following administration of exogenous stem/progenitor cells (De Santis et al., 2018). This involves a cautious, but expanding, examination of cell treatments in lung illnesses in clinical trials. However, alveolarization processes must first be studied to discover signaling mediators that may be used to induce alveolarization in the injured lung. Recently, several potential cell types, pathways, and molecular mediators have been discovered. Using lineage tracing methods and lung damage models, novel progenitor cells for mesenchymal and epithelial cell types and cell lineages that can acquire stem cell attributes have been discovered.

2. Lung development

A comprehensive knowledge of lung development during gestation periods is essential when discussing lung regeneration and disease models.

Much of our knowledge about lung development comes significantly from mouse lung developmental research. Even though there are distinctions between the mouse and human lungs, the mouse genetic models remain the most robust *in vivo* model for studying the fundamentals of lung development. There are three stages in the development of the mammalian fetal lung: (1) first outpouching of the endoderm tube that develops the lungs, and initial differentiation in the anterior foregut of the lung endoderm and mesoderm progenitors occurs; (2) hundreds to thousands of distal ends form as a result of branching morphogenesis, by arborizing this rudimentary lung tube in a typical manner (pseudo-glandular stage); (3) the last phase of lung development culminating in sacculated stage and alveolarization resulting in the formation of operational gas exchange units required for postnatal respiration (Morrisey & Hogan, 2010; Nikolić, Sun, & Rawlins, 2018; Schittny, 2017).

Amid the 7th and 16th weeks of embryonic development, the pseudo-glandular stage occurs, progressive branching generates conducting airways, and rudimental airways are developed over the course of 16–25 generations. Only when endodermal lung buds are exposed to bronchial mesoderm can they branch. The amount of mesenchyme present appears to be closely correlated to the rate and breadth of branching. By 16 weeks, all bronchial airways have developed. After this period, existing airways elongate and expand, resulting in additional expansion. The initial differentiation of lung epithelium occurs at this Stage (Woods, Schittny, Jobe, Whitsett, & Abman, 2016). Cilia emerge in the proximal airways by 13 weeks. For this epithelial differentiation to occur, the mesenchyme plays a vital role, resulting in a shift from ciliated columnar and goblet cells or bronchial epithelial cells to alveolar type II cells.

On the other hand, lung epithelium is required for lung mesenchyme differentiation. The canalicular stage lasts from the 16th to the 25th week. The gas-exchanging component of the lung is produced and vascularized at this Stage (Herriges & Morrisey, 2014; Schittny, 2017). As the interstitial tissue diminishes, the capillary network is observed to expand. By week 20, alveolar type I pneumocytes (AT1) are the primary structural cell of the alveolus. The type I pneumocytes are large, flattened, and entirely differentiated, and the exchange of gas occurs across the extremely thin membranes of these cells. Capillaries start to develop and appear in close proximity to the alveolar cell's distal surface (Harding & Hooper, 1996; Nikolić et al., 2018; Schittny, 2017). Around the same period, lamellar bodies, also known as inclusion bodies, begin to form in alveolar type II cells (AT2). These lamellar bodies function to store surfactant, which later is released into the alveolar space.

The terminal sac also referred to as the saccular stage, generally lasts from 26 weeks until term. There is a reduction in the interstitial tissue and a weakening of the airways (alveolar) walls at this stage, Type I and Type II pneumocytes are identifiable as this stage progresses. At this stage, a phosphatidylinositol-rich surfactant is required for alveolar stability vs phosphatidylglycerol and phosphatidylcholine essential in late gestation lungs. The quantity of lamellar bodies present at birth correlates to the stability of the lungs.

Without surfactant, the lung can only keep alveoli open for a limited period of time. There are roughly 20×10^6 saccules at birth, and the air-containing area, eventually becoming the alveolus, is referred to as a "primitive saccule" during birth. Following the lung's histological and functional development, the saccules continue to mature. The respiratory bronchiole is lined by cuboidal cells, and the canalicular phase lasts from the 16th to the 26th week of pregnancy. The terminal sac phase is initiated by the 6th and the beginning of the 7th prenatal month. While mature Type I cells line these saccules, the form or geometry of the saccules does not reach "adult" morphology until around 5 weeks post-birth (Surate Solaligue, Rodríguez-Castillo, Ahlbrecht, & Morty, 2017). The functional alveolus contains surfactant generated by Type II cells and is connected to an alveolar duct, which in turn, in close proximity to pulmonary capillaries, is lined with Type I cells and has pores (Kohn pores) linking it to neighboring alveoli (Whitsett, 2014). Simultaneously two alveoli are exposed to the interstitial capillaries. The air/blood contact barrier is formed by the type I cell, a very thin basement membrane, and the pulmonary capillary endothelium. The lung has around 300×10^6 alveoli at functional maturity. By the age of eight, this number of alveoli appears to have been attained (Nikolić et al., 2018). The oxygen and carbon dioxide exchange between air and blood in the human lungs is facilitated by very thin tissues made up of AT1 pneumocytes, which are exceedingly delicate and vulnerable to damage.

3. Lung stem cell diversity

The lungs are exposed to myriad sources of pollutants, toxins, and other environmental hazards. Normal lung shows a slow turnover of cells but once injured, lung stem cells show high self-renewal, proliferation, and subsequent differentiation to maintain cellular homeostasis (Paul et al., 2014). In cases of major injuries like surgical removal of a lung lobe, extensive injury repair, and compensatory neoalveolarization occurs (Hoffman et al., 2010). A reservoir of adult stem cells (ASCs) contributes to the repair and regeneration of the lungs in the event of an injury. Stem cells are cells with varying potency levels and the ability to self-renew. They serve an essential role in the regeneration of organs, repair, and maintenance. Putative stem/progenitor cell types and their associated niches are found in the different subsections of the adult lung (trachea, bronchi, bronchioles, and alveoli) and yield distinct differentiated cell types. The region-specific stem/progenitor cell types that restore epithelial barrier functions after injury are shown in Fig. 1 and will be discussed in this section. Many of these niches in the intricate microscopic structure of the lungs are inhabited by epithelial progenitor cells. It has been demonstrated that multipotent endogenous stem cells exist in completely matured lungs, and these cells live in niches whose microenvironment controls their ability to proliferate and differentiate in the wounded organ (Ratajczak et al., 2017; Whitsett, 2014).

The lungs are slow-turnover organs with a great potential to restore epithelial damage after acute traumas, and various highly plastic stem cell populations (Lee & Rawlins, 2018) will be discussed in order of the proximal-distal axis. The submucosal glands (SMGs) are glands that release serous fluids and mucus and moisturize and disinfect the airways. Antimicrobials such as lysozyme, lactoferrin, and lactoperoxidase are found in the serous fluid generated by SMGs. SMGs line the airways and are embedded under the connective tissue between the surface epithelium and cartilage (Hegab et al., 2012). The SMGs are anatomically found only in the proximal section of the mice trachea; however, the SMGs line across the trachea and bronchi in higher mammals, including humans. Airway basal stem cells (ABSCs) (Tp63+, Krt5+, PDPN+, NGFR+, Krt14+) are the major ASCs in the airway surface epithelium and SMGs that have tremendous regeneration and differentiation capacity and play a vital role in the repair and regeneration of the epithelial layer (Pardo-Saganta et al., 2015; Paul et al., 2014; Rock et al., 2009; Rock, Randell, & Hogan, 2010). The ABSCs are localized in the trachea and extend to the main stem bronchi in mice, but ABSCs extend up to the intermediate airways in humans (Kotton & Morrisey, 2014; Rock et al., 2009, 2010). The multipotent basal stem cells are housed near the basal lamina and at the base of the pseudostratified mucociliary epithelium and can differentiate into all the airway cell types, including the ciliated cells (FoxJ1+, CCDC40+), goblet cells (submucosal MUC5B+, airway MUC5AC+), club cells (SCGB1A1+), neuroendocrine (NE) cells (CHGA+, CGRP+),

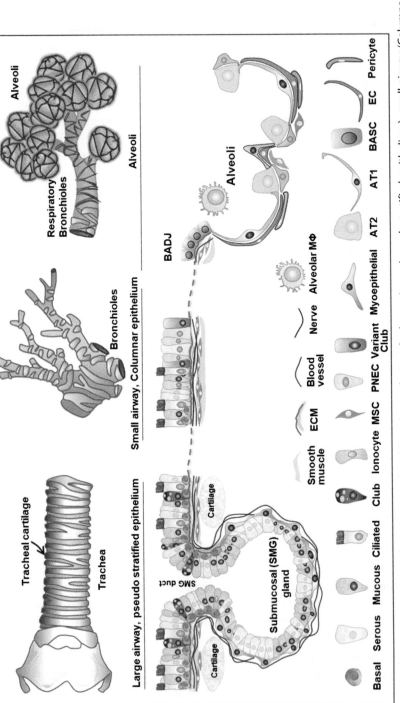

Fig. 1 Adult lung stem/progenitor cells and their niches. The scheme depicts the large airway (pseudostratified epithelium), small airway (Columnar epithelium), respiratory bronchioles, and alveoli. Proximal to distal axis-wise spatially distribution of prospective lung epithelial stem/progenitor cells, as well as the differentiated pulmonary cells. Along the Proximal to the distal axis of the airway, several region-specific potential stem cell niches exist. The potential progenitor/stem cells are found in their own local niches, where they retain their stem/progenitor qualities and the ability to develop into diverse lung cell types. SMG, submucosal gland; BADJ, bronchoalveolar-duct junction; PNEC, Pulmonary neuroendocrine cell; MSC, Mesenchymal stem ~~cell; AT1 Alveolar type I cell; AT2, Alveolar type II cell; EC, Endothelial cell; BASC, Bronchoalveolar stem cell; MΦ, Macrophage; ECM, Extracellular matrix.~~

ionocytes (FoxI1 +, Ctrf+), and tuft cells (Pouf2f2 +, Trpm5 +) (Busch, Lorenzana, & Ryan, 2021; Montoro et al., 2018).

New transcriptome evidence points to the existence of ABSC subpopulations, making the situation more complex (Carraro et al., 2020; Smirnova et al., 2016; Watson et al., 2015). An ABSC subpopulation recently discovered is the "hillock" (Krt13 +) cells and may be associated with the formation of squamous epithelium (Montoro et al., 2018). Other variants of the basal stem cells (Trp63 +, Krt5 +) are also described as distal alveolar stem cells (DASCs) and are a cluster of cells that may replace injured alveolar cells (Kumar et al., 2011). While other researchers postulated a Trp63 +, Krt5 + population as lineage negative epithelial precursor (LNEP) cells that help regenerate alveoli damaged by bleomycin (Vaughan et al., 2014). There are also progenitor cells that can play important roles in airway regeneration. The ABSCs may also give rise to intermediate luminal progenitor (Krt8 +) cells with limited differentiation and proliferation potential. The SMG duct basal stem cells may differ from ABSCs (Hegab et al., 2014). More research is needed to pinpoint the functional distinctions between the SMG duct basal stem cells, ABSC subpopulations, progenitor subpopulation, and their relevance in humans and other animals in the context of development and disease.

Small airways lack ABSCs and are lined with cuboidal epithelium (with secretory, ciliated, and clustered neuroendocrine cells), leading to the broncho-alveolar duct junction (BADJ) bordering the alveoli. The multipotent bronchoalveolar stem cells (BASCs) (CC10+, SPC+) are found in the BADJ and can differentiate into both bronchiolar epithelium (including club and ciliated cells) and AT2 and AT1 cells for distal airway repair (Bertoncello, 2016; Donne, Lechner, & Rock, 2015; Liu et al., 2020; Ratajczak et al., 2017). They multiply in response to injury in the airways and the alveoli. BASCs have been identified in mice, but their role in humans has yet to be thoroughly investigated (Kim et al., 2005). Airway damage stimulates BASCs, resulting in a rise in the tumor number and size, suggesting that BASCs are involved in lung cancer (Kim et al., 2005; Yin et al., 2022). The adult lung includes mature epithelial and mesenchymal cell types with no embryonic progenitor populations remaining. Many lung cells can become plastic during restoration, with the extent of the damage apparently deciding which cells will eventually respond (Tata & Rajagopal, 2017). Although this plasticity provides the lung with flexible, strong, restoring mechanisms, it has sparked discussion and perplexity among stem cell researchers. However, several questions remain unanswered, like what prevents club cells from transforming into the basal cells in the basal cell-deficient bronchioles. How are the different stem cell populations restricted to different zones, and what is their putative niche? What is the relevance of these different stem/progenitor cell populations in different lung diseases?

A cluster of pulmonary neuroendocrine cells (PNECs) cells, usually located at the bifurcations of the airway, are termed neuroepithelial bodies (NEBs). PNECs typically work as intrapulmonary sensors and are reported to operate as progenitors during injury-induced healing. Researchers used the mice naphthalene-induced club cell loss model and found that some NEBs (CGRP+, SCGB1A1+) may lead to club and ciliated cells (Reynolds, Giangreco, Power, & Stripp, 2000; Song et al., 2012). Recent studies found that epithelial damage increased Notch signaling in PNECs (Yao et al., 2018). An injured PNEC with Notch2 may dedifferentiate and subsequently re-differentiate into other airway epithelial cells, according to scRNA-seq (Ouadah et al., 2019; Xu, Yu, & Sun, 2020). Leucine-rich repeat-containing G-protein-coupled Receptors (LGR) 6 (LGR6+) cells in the bronchioalveolar compartment of the human lung have also been reported to demonstrate the capacity to self-renew and expand indefinitely, as well as express stem cell markers (SOX9, LGR5–6, ITGA6) (Cortesi & Ventura, 2019). *In vivo* investigations revealed that the LGR6+ cells also have the capacity to develop multiple bronchioalveolar mature cell types (Club cells, AT1, and AT2 cells) and regenerate damaged tissue (Oeztuerk-Winder, Guinot, Ochalek, & Ventura, 2012).

Distally the AT2 (Sftpc+) cells are specialized stem cells found in lung tissues that may replace damaged AT1 cells (HOPX+, AQP5+) and ensure lung cell regeneration. AT2 cells, which reside in the corners of alveoli and produce a high amount of surfactant protein C, are well-known lung stem cells (Barkauskas et al., 2013; Donne et al., 2015; Ratajczak et al., 2017). According to several study findings, AT1 cells are somatic cells present in the lungs and lack stem cell attributes, these are cells derived from AT2 cells. The basic understanding is that almost 95% of the interior surface of each alveolus is covered with AT1 cells, which are squamous cells, thin and optimal for gas exchange (Li, He, Wei, Cho, & Liu, 2015). They share a basement membrane with the endothelium of pulmonary capillaries, forming the air-blood barrier across which gas exchange takes place. Connecting *via* tight junctions, these pneumocytes and other alveolar cells provide an impenetrable barrier to prevent fluid penetration into the alveoli (Brandt & Mandiga, 2021). Recently, Hopx+ AT1 cells were reported to show plasticity following pneumonectomy. AT2 cells proliferate and differentiate to rebuild the alveoli through a TGF signaling pathway (Jain et al., 2015). These findings imply a bi-directional transdifferentiation potential of mature alveolar cells, especially during severe injury conditions and pneumonectomy. A precise analysis of all lung stem/progenitor cell populations is needed to distinguish their actual location, markers, progeny, and function at the single-cell resolution.

4. Lung regeneration

Novel gene editing technologies and *in vitro* models of pluripotent stem cells have provided models for studying lung disease, investigating lung regeneration processes, and raising the hope of repairing lung malfunction. These insights into the biology of long development are currently being used to understand the processes that drive the regeneration of lungs. We will go through how fundamental research is now being applied to regenerative therapies, opening new research pathways that might eventually lead to lung disease therapies (Stabler & Morrisey, 2016). It has been observed that acute epithelial injury in the lungs can trigger a robust regeneration response fueled by proliferative responses from healthy nearby epithelial cells (Aros, Vijayaraj, et al., 2020; Rock et al., 2009, 2010). Lung regeneration after acute damage involves numerous progenitor lineages; therefore, depending on the geographical niches involved, this process might be thought of as following a principle that is democratic (Kotton & Morrisey, 2014; Paul et al., 2014). The basement membrane beneath the epithelial and endothelial cell layers is considered to be essential in promoting stem cell migration into the injured area by providing a scaffold and the ability to concentrate paracrine secreted bioactive chemicals during successful lung regeneration (Beronja & Fuchs, 2011; Desai, Brownfield, & Krasnow, 2014; Hegab et al., 2014).

The signaling mechanisms employed by these "progenitors" are more commonly the same as those used in developing lungs (Akram, Patel, Spiteri, & Forsyth, 2016; Bertoncello, 2016; Paul et al., 2014; Rock et al., 2010). The activation and subsequent differentiation of Trp63 + basal cells into secretory and multiciliated epithelial cells in the central airways and trachea are regulated by a complex interplay of Notch signaling, highlighting the role of this pathway in airway epithelial development (Mori et al., 2015; Rock et al., 2009, 2010; Tata & Rajagopal, 2017). Notch-3 inhibits basal cell overexpansion in the conducting airways (Mori et al., 2015;

Pardo-Saganta et al., 2015), while Notch-2 activation is required for alveologenesis in the alveoli (Pardo-Saganta, Tata, et al., 2015; Tsao et al., 2016). Recent research has also shown that some lung airway epithelial cells demonstrated unique flexibility after significant damage, as secretory cells dedifferentiated back into basal cells to restore the basal cell population (Tata et al., 2013). The extent to which this plasticity plays a role in physiological regeneration and the molecular mechanisms governing it are major critical questions that need further investigation. The formation of a BASC lineage in the more distal lung airways has been discovered to be dependent on Wnt signaling, which is regulated by Gata6, a key transcription factor in lung development (Kim et al., 2005; Zhang et al., 2008). BASC formation is influenced by BMP4 expression in the lung endothelium, which promotes TSP-1 expression, and ultimately modulates BASC differentiation (Lee et al., 2014). During regeneration, these developmentally conserved signaling pathways are located not just in the epithelium but also in the supporting mesenchymal lineages. Another signaling pathway known as Hedgehog signaling promotes mesenchymal quiescence in the adult lung, and the inhibition of hedgehog can result in enhanced epithelial proliferation during homeostasis and after damage (Peng et al., 2015).

Lung disease and regeneration may be studied using a number of models, including in vivo and in vitro models. The mouse in vivo model is the most commonly utilized in lung regeneration studies. While there are substantial biological and physiological differences between the mouse and human lung, recent findings in human lung research show that most, if not all, of the significant cell lineages are similar, and many of the essential transcriptional and signaling pathways are conserved. Chiefly, the discovery of these key regulators of lung development and regeneration in mice has paved the way for groundbreaking new research in humans, including the use of primary human lung cells in lung organoid culture systems and the generation of human lung epithelial cells such as AT2 cells from pluripotent stem cells (Logan & Desai, 2015; Longmire et al., 2012; Mou et al., 2012; Rock et al., 2009, 2010). Initials attempts to generate lung epithelial lineages from mouse or human embryonic stem cells (ESCs) were inefficient or unsuccessful until it was revealed that activin A activation could force ESCs to recapitulate early embryonic development, resulting in efficient definitive endoderm (Ishizawa et al., 2004). Subsequently, other researchers have used growth factors to guide the development of iPSCs to lung epithelial cells and ESCs with varying degrees of success since then (Ghaedi et al., 2017; Gomperts, 2014; Wilkinson et al., 2017).

Research by Schiller and colleagues demonstrates the relevance of tissue remodeling in the regeneration process (Schiller et al., 2015). Their work studied the dynamics of extracellular components evaluated from the moment of lung injury induced by bleomycin through transitory fibrotic healing by day 14 and lung regeneration from the 4th to 8th week. During the first week of exposure to bleomycin, a bacterially produced chemical, death of a variety of epithelial cell types, including Club cells, endothelial cells, and AEC1, AEC II, many changes occur in the residual tissue as a result of bleomycin-induced cell destruction, and some extracellular matrix proteins, such as Emilin2, which interacts with elastin, become significantly elevated. From research done previously, signal transduction pathways and matrix composition influence the rise in proliferation following partial pneumonectomy (PNX) (Vijayaraj et al., 2019; Voswinckel, 2004). The concentration of elastin in the body, for example, is thought to have a function in regulating the proper growth and structure of the alveolar compartment (Hilgendorff et al., 2015). Retinoic acid, HGF, FGF, and EGF signaling have all enhanced alveolar regeneration post-PNX (Olajuyin, Zhang, & Ji, 2019; Panganiban & Day, 2010; Rabata, Fedr, Soucek, Hampl, & Koledova, 2020). Greater attention to the mechanisms that promote endogenous lung regeneration and the various stem/progenitor lineages and geographic niches involved would bring the field a step closer to developing novel pharmacological methods for the repair of human lungs and chronic disease therapies.

5. Lung stem cell niche

A stem-cell niche is an anatomic tissue section that offers an intricate and adaptable specialized milieu where stem cells can exist in an undifferentiated, self-renewing form. The stem-cell niche or microenvironment is mainly composed of extracellular matrix (ECM) bioscaffold, adhesion-associated proteins, soluble/insoluble signaling molecules, and cells that interact with and harbor stem cells to modulate their function especially proliferation, differentiation, and stem cells quiescence. These factors may come from nearby epithelial (stem, progenitor, and differentiated cells), stromal (mesenchymal cells, fibroblasts, smooth muscle cells, and endothelium), and immune cells. We provide evidence for each of them as niche components for lung epithelial stem cells. The niche can be regulated by juxtacrine, endocrine, paracrine, and autocrine signals and might be spatiotemporally distinct. Niche interaction with the stem/progenitor cell's intrinsic transcriptional regulatory network guides long-term stem cell maintenance, lineage-specific differentiation, and tissue homeostasis following lung injury. Post-injury, the lung stem cells must produce an appropriate ratio of stem, progenitor, and differentiated cell types to maintain cellular homeostasis and restore lung function. Another important aspect is to arrest regeneration post-repair once the structure and function of the tissue are restored to avoid undesired stem cell proliferation that might lead to lung diseases. The following section will describe the anatomically distinct putative stem/progenitor cells and their niches as oriented in the proximal–distal axis.

5.1 Early development of the lung and niche

The molecular mechanism regulating developmental competence and lung development is dependent on the endoderm–mesoderm interaction. During development, proximal-distal lung patterning is characterized by restricting the distal lung epithelial progenitors from diffusing into the proximal epithelial cells. This is primarily regulated by the fibroblast growth factor 10 (Fgf10) secreted by the distal mesenchymal cells (El Agha et al., 2014). Fgf10 restricts the fate to distal progenitors (Sox9+) and prevents them from acquiring bronchiolar (Sox2+) fate. Fgf10 remains coupled with heparin sulfate (HS) in the basement membrane of distal progenitors. Thus, the cell surface associated with HS chains helps transmit Fgf signals to nearby cells (Volckaert & De Langhe, 2014). Many developmental signals are used during injury repair in adult lungs and are more interesting considering the disease perspective. In this chapter, we have mainly focused on the adult lung stem cell niche, and the developmental aspect of the lung is discussed in details in other review papers (Schittny, 2017; Varghese, Ling, & Ren, 2022).

5.2 Proximal airway stem cell niche in adult lung

The proximal airways are well studied in mice as they show a strong resemblance to the human trachea. The activation of stromal cells, mainly fibroblasts, to produce paracrine signals is critical for niche formation in stem cell homeostasis. The stromal niche-mediated molecular cues influence proliferation, self-renewal, differentiation of stem cells, and maintenance of tissue homeostasis post-injury. Tight regulation of these feedback loops is critical to maintaining a normal tissue state, and an aberration is related to diseases. The airway repair process involves two different steps. After epithelial denudation, ABSCs undergo symmetric (self-renewal) and asymmetric division (giving rise to Krt8 + early progenitors) (Paul et al., 2014; Rock et al., 2009; Tata et al., 2013). After the proliferative phase, differentiation sets in to reform the pseudostratified epithelium. Among the several niche-associated signaling pathways, Wnt signaling also plays a vital role in the airways. Wnt signaling works in close range, and Wnt ligands are secreted from a cell when PORCN, a transmembrane O-acyltransferase acylates and promotes its secretion (Chatterjee et al., 2022). Aros et al. in 2020 used various transgenic mice models to investigate how Wnt signaling influences proximal ABSC homeostasis, injury healing, and aging. Data suggested the existence of distinct spatiotemporally dynamic Wnt-secreting niches that control ABSC fate, airway regeneration, and homeostasis. Post-injury inter-cartligeneous zone (ICZ)-restricted stromal Pdgfra + fibroblasts transiently active PORCN and secreted Wnt for *in vivo* proximal airway epithelial self-renewal and proliferation (Aros, Paul, et al., 2020; Aros, Vijayaraj, et al., 2020). As the cells tend to differentiate, the fibroblastic niche gives way to the intraepithelial niche that guides differentiation to ciliated cells *in vivo* because K5 + ABSCs secrete Wnt ligand. More experiments are needed to better understand the role of the Wnt secreting niches in different injury models and disease states.

A stem cell may undergo a symmetric division for self-renewal and undertake an asymmetric division to form progenitor cells in the airway epithelium (Paul et al., 2014). A stem/progenitor cell may also act as a niche for its descendants, and this forward signal is required for daughter cell maintenance. ABSCs continuously provide Notch ligand to their offspring progenitor cell and thereby maintain a secretory cell fate, while an absence of this stem cell-derived signaling means a terminal differentiation toward a ciliated cell fate occurs (Pardo-Saganta, Tata, et al., 2015). We are still learning how the proximal airway stem cell niche interacts and maintain stem cell homeostatic, regenerative, and adaptive actions. Next, we describe the regulatory niches for distal lung epithelial stem cells.

5.3 Distal stem cell niche in adult lung

Lung development is a multistep process involving various cell types, factors, and mediators. Experiments using fate-mapping, single-cell sequencing, and advanced microscopy have focused on unique cell types that may assist the pulmonary alveoli in self-renewal and repair.

5.3.1 Mesenchymal cells and lung stem/progenitor cell niches

Several developmental pathways are rekindled during the injury-damage response in adults. Recent data suggest that airway smooth muscle cells (ASMCs) serves as a club cell niche (Volckaert et al., 2011). Researchers using a naphthalene-induced epithelial damage model found injury-induced epithelial secretion of Wnt7a, which activated ASMC and stimulated the Fgf10 production. This niche signaling stimulates the progenitor function of the club cell, leading to club cell proliferation and subsequent epithelial repair at the BADJ area. Fgf10 overexpression led to an increased epithelial repair, while inhibition of the signaling impaired the regenerative potential of club cells after naphthalene-induced epithelial damage (Gupte et al., 2009; Volckaert et al., 2011). The expression of the Fgf10 by ASMCs has been reported with various airway epithelial injuries, including that of ozone and bleomycin. It is interesting to note how developmental signaling cascades can play a role in epithelial repair and regeneration in the adult stem/progenitor cells (Volckaert & De Langhe, 2014). A recent study using single-cell RNA sequencing elucidated the role of Fgf10-Fgfr2b signaling in the generation of the bronchial neobasal cell and the activation of AT2 cells driving alveolar epithelial repair and regeneration following bleomycin-mediated injury (Yuan et al., 2019). Bleomycin-mediated injury leads to an expansion of Fgf10-expressing fibroblasts, further strengthening the role of Fgf10 as niche signaling. Fgf10 is highly expressed by the mesenchymal cells in SMG ducts and the intercartilaginous zones of the airway and may serve as niche signaling for the SMG duct basal stem cells and ABSCs (Sala et al., 2011; Tiozzo et al., 2009; Yuan, Volckaert, Chanda, Thannickal, & De Langhe, 2018). Using a 3D organoid model, a recent study exhibited that Fgf signaling activation in organoids led to alveolar-like regions (Yuan et al., 2019). More studies are needed to investigate the role of Fgf10 is the niche signaling for other lung stem/progenitor cells.

BASCs have the potential to generate both the bronchiolar and alveolar differentiation and hence differential niche mediated signaling may be regulating the directionality post-injury. Experimental evidence suggests that lung endothelial cells can deliver niche signaling to induce differentiation toward the alveolar lineage (Lee et al., 2014). Lung endothelial cell–derived Thrombospondin-1 (Tsp1) stimulates the differentiation of BASCs into the alveolar lineage. After the bleomycin-mediated injury, BASCs release Bmp4, which works on endothelial cells and, through a Bmpr1a/calcineurin/NFATc1-dependent pathway, activates the production of Tsp1 (Lee et al., 2014). This is interesting as Bmp4 and Tsp1 expression is decreased following a naphthalene-mediated injury where the club cells are mainly damaged, suggesting the role of Tsp1-mediated signaling in the directionality of differentiation (Van Winkle, Buckpitt, Nishio, Isaac, & Plopper, 1995). Bmp4 and Fgf10 signaling crosstalk occur during lung development, and the role of Fgf10 in regulating Tsp1-mediated niche signaling needs further investigation after different types of epithelial injury. As BASCs can respond to both bronchiolar and alveolar injury, activating signals from both the proximal and distal epithelial layers may activate BASCs in the BAJ niche, allowing them to develop uni/bi-directionally into proximal bronchiole or distal alveolar cells. Fibroblasts are reported to act as a niche by secreting TNF, which stimulates the Lgr6 + stem/progenitor cell subpopulation during injury. A paracrine feedback loop is initiated *via* p38α-SDF-1 is released by Lgr6 + stem cells, which stimulates the fibroblast to secrete niche signaling.

Ascl1-expressing NEB cells may act as a niche for a subpopulation of Scgb1a1 + club cells in the developing airways. This Clara-like secretory precursor subpopulation (SCgb3a2 +, Upk3a +) remains adjacent to NEBs and contributes to the club (Scgb1a1 +) and ciliated cells in the adult lung (Guha et al., 2012). Though it appears that pulmonary NEBs constitute a stem cell niche necessary for club cell regeneration after lung damage, lungs devoid of NEBs recover properly post naphthalene-mediated injury. Suggesting that further investigation is needed to dissect the role and mechanism of NEBs as a niche. Recent work sheds light on the formation and maintenance of the human Lgr6 + alveolar epithelial stem cell niche, critical for stem cell self-renewal potential. The scientists discovered a paracrine circuit that attracts and stimulates fibroblasts to release TNF. TNF subsequently activates an autocrine loop mediated by TGF/p38 in Lgr6 + stem cells, further increasing SDF-1 synthesis. These elevated SDF-1 levels subsequently induce fibroblasts to create angiogenesis-promoting factors (Ruiz, Oeztuerk-Winder, & Ventura, 2014).

Myofibroblasts and lipofibroblasts (LIF) are two types of interstitial fibroblasts in the developing lung. While the myofibroblasts undergo apoptosis following alveolarization, lipofibroblasts (PPAR-16 +) transport lipids and help the AT2 cells in the formation of pulmonary surfactant (Chen, Acciani, Le Cras, Lutzko, & Perl, 2012; Donne et al., 2015). LIFs in the adult lung parenchyma are present apposed to the AT2 cells and are hypothesized to maintain their stemness and act as AT2 progenitor cell niche (Chen et al., 2012). In the developmental stages, the Fgf10-ERK1/2 pathway acts on the distal progenitor cells and restricts them to an AT2 fate (Li et al., 2018).

This developmental pathway is restored during injury. Fgf10 is secreted by LIFs during bleomycin-induced lung fibrosis and is known to directly activate SFTPC expression in AT2 cells. Fgf10 activates proliferation and also protects the AT2 cells from injury-induced apoptosis. In *in vitro* culture conditions, clonal growth of Sftpc+ AT2 stem cells (ITGA6+, ITGA4+) requires coculture with Sca1+ mesenchymal cells that may be replaced by exogenous Fgf10, suggesting the role of paracrine Fgf10 as niche-mediated signal (McQualter, Yuen, Williams, & Bertoncello, 2010).

Not only the Fgf10 signaling is essential, but the role of Wnt signaling is also important in niche-mediated signaling. Pdgfra-expressing stromal fibroblasts reside in close proximity with the AT2 cells and showed high expression of PORCN. Many of these cells are even single cells and secrete Wnt ligands, thereby activating Wnt signaling in AT2 stem cells, maintaining stem cells, and inhibiting transdifferentiation to AT1 cells. Interestingly, AT2 autocrine Wnts are induced in case of an injury, leading to the proliferation of the AT2 progenitor cell pool. Interestingly a loss of this single-cell niche leads to a lack of the juxtacrine Wnts and loss of AT2 stem cells, while during alveolar epithelial damage, AT2 cells produce autocrine Wnts that transiently enlarge the progenitor pool (Nabhan, Brownfield, Harbury, Krasnow, & Desai, 2018). Thus, a dynamic shift happens in the niche to restore alveolar homeostasis, somewhat similar to the ABSCs (Aros, Vijayaraj, et al., 2020).

5.3.2 Immune cells and lung stem/progenitor cell niches

Recent data have established that other than epithelial or mesenchymal cells, the recruited and resident immune cells are essential members of stem cell niches throughout the body and play their role in the reparative response. In the lung, resident immune cells are critical in providing local immunity and protection from infection, especially during lung tissue injury (Kathiriya & Peng, 2021). Thus, resident and trafficked immune cells interact with the stem cells and act as a niche in a context-specific manner. Injury and inflammation attract immune cells, and thereby, they traffic into the stem cell niche and impact the regenerative outcome. Following viral infection, infiltrating immune cell-secreted IL-1/TNF can induce signaling through NF-κB in the AT2 cells (SFTPC+) and promotes proliferation and subsequent differentiation into AT1 cells (Katsura, Kobayashi, Tata, & Hogan, 2019). Partial pneumonectomy is the process of removal of one or more lung lobes, and in both mice and humans, a compensatory development of the remaining lung lobes is observed post-surgery. A surge in stem cell proliferation and tissue regeneration reestablish sufficient gas exchange capability and lung function. Researchers used a unilateral pneumonectomy model and observed CD115+, and CCR2+ monocytes and M2-like macrophages infiltration coincided with the surge in AT2 cell proliferation. Also, bone marrow-derived macrophages (BMDMs) and Il4ra-expressing leukocytes infiltrate the lung and play a role in lung regeneration (Lechner et al., 2017). Thus, injury-associated inflammatory responses can play a crucial role in lung regeneration (Verheyden & Sun, 2020). Another study suggested the role of interstitial macrophage-secreted IL-1β as an inflammatory response, activates the generation and proliferation of a subset of AT2 cells (Il1r1 +) into a transitional progenitor state or called damage-associated transient progenitors (DATPs). The immune-mediated signaling also induces the differentiation of AT2 cells into AT1 cells (Choi et al., 2020). Following this study, Choi, Jang, and colleagues demonstrate that immune cell and differentiated epithelial cell interaction can modulate the niche signaling circuitry, leading to stem cell fate alteration following bleomycin injury. Immune cells mediated inflammatory signals reduce the expression of the ciliated cell (Il1r1 +) specific Notch ligands (Jag1/2), which are in direct contact with distal airway stem cells. IL-1 secreted by immune cells induced decreased notch signaling in the ciliated cell, which eventually resulted in the activation of the Fra2/Fosl2 transcription program in airway stem cells, facilitating alveolar fate. In this instance, a differentiated cell acts as a niche by sensing the change in the immune repertoire and relays this information to the stem cells, thereby regulating the fate (Choi et al., 2021).

5.4 Niche and lung diseases

Repair of the airway epithelium by stem cells is normal, but an aberrant repair may lead to lung diseases, especially lung fibrosis and cancer. The lung's stem cell niche is likely crucial in deciding on an active and normal repair process (Gomperts & Strieter, 2007). Limited efforts are made to dissect the molecular nuances of a diseased lung niche or disease-associated altered niche. The genetic alteration, repeated injury, inflammatory microenvironment, and aging can alter stem/progenitor cell niche and regenerative potential leading to chronic and acute lung disease (Melo-Narváez, Stegmayr, Wagner, & Lehmann, 2020). Another aspect is that the development of a diseased niche may lead to attenuated stem/progenitor cell plasticity and regeneration. Lung diseases like Chronic Obstructive Pulmonary Disease (COPD) develop due to repeated insults and associated chronic inflammation in the distal airways and alveolar niche (Barnes, 2016). Recent COPD patient single-cell RNA-sequencing (scRNAseq) data of the alveolar niche cells was performed to decipher a diseased niche and the mechanism of disease. Data suggested alteration in the endothelial cells with high chemokine secretion as a cause of inflammation in COPD, while a metallothionein + alveolar macrophages subpopulation constitutes a disease niche (Sauler et al., 2022). Strong indications are evident of an altered lung stem/progenitor cell niche, as an increase in NEB is reported in multiple diseases like diffuse idiopathic pulmonary neuroendocrine cell hyperplasia (DIPNECH), neuroendocrine hyperplasia of infancy (NEHI), asthma, and bronchopulmonary dysplasia (BPD) (Little et al., 2020; Sui et al., 2018). The aberrant amplification of NE cells most likely alters the microenvironment and the niche leading to an increase in the NEB-associated clusters of CCSP-expressing cells in BPD lungs (Noguchi, Furukawa, & Morimoto, 2020).

Bleomycin-associated fibrosis is associated with a subset of myofibroblasts (Axin2 +) forming the myofibroblast niche, which is fibrogenic and pathologically adverse, characterized by rapid proliferation and ECM component deposition (Zepp et al., 2017). The balance of interacting homeostatic signals emanating from stem/progenitor cells, stromal, and immune cells may be a key factor determining the appropriate regeneration post-injury. A failure to maintain the proper balance may result in pathological processes (e.g., lung fibrosis, cancer metastasis), in which inflammatory signaling promotes stromal compartment growth while impairing epithelial differentiation and effective tissue repair. Thus, further studies to better understand the mechanisms and consequences of the cellular interactions in this microenvironment have potentially high clinical significance and may lead to better-targeted therapeutics. The creation of artificial niches can play a crucial role in dissecting the molecular intricacies of niche-stem/progenitor cell interaction.

6. Recapitulating stem cell niche

The human body's stem cells are unspecialized and have the capability to renew themselves and can mature into any type of cell. On average, stem cells are mostly a self-maintaining group if each stem cell differentiation produces one substitute stem cell and one transit-amplifying cell. They represent a small percentage of overall proliferative cells and are basically undifferentiated. The functional capabilities of the progeny that stem cells inevitably lead to are not present in most tissues. Stem cells cycle slowly and yet are highly clonogenic. Adult and embryonic stem cells exist (Zakrzewski, Dobrzyński, Szymonowicz, & Rybak, 2019). Stem cells exist in a stem cell niche, indeterminate and self-renewable form. This tissue section supplies the cells with a specialized habitat to dwell in (Birbrair & Frenette, 2016; Donne et al., 2015). Schofield was the first to conceptualize a stem cell niche as a milieu dedicated to stem cells and proposed that this niche may provide an anatomical region that might regulate the population of stem cells retained (Pajonk & Vlashi, 2013). The stem cell niche comprises a subpopulation of cells in a particular region in charge of monitoring committed stem cells that can sustain themselves. The niche might vary greatly; however, it is a physical anchorage for stem cells and has complete control over their functioning (Krause, Conboy, & Conboy, 2017).

For a growing fetus, copious niche factors impact embryonic stem cells throughout embryogenesis to modify the expression of genes and drive proliferation or differentiation. Adult stem cells are maintained in an inactive form within the human body by stem-cell niches, but in case of a tissue is injured, the adjoining region dynamically stimulates stem cells to encourage regeneration to develop new tissues (Jhala & Vasita, 2015). The stem cell niche includes a spatial arrangement contributing to stem cell fate determination and supporting existing clones through morphological and physiological interactions (Pennings, Liu, & Qian, 2018). Interactions involving two stem cells as well as a stem cell and adjoining specialized cells, integrins, ECM components, osmotic pressure, cofactors, cytokines, and the physiochemical properties of the niche are all influential in determining stem-cell attributes within the niche (Jhala & Vasita, 2015).

Adult human stem cell niches are far more intricate due to the accessibility of many possible stem cell locations and many prospective niche cells and extracellular elements implicated in regulating and maintaining tissue-specific stem cells (Walker, Patel, & Stappenbeck, 2009). Neural, hematopoietic, bone marrow, intestinal, and epithelial stem cell niches, in addition to the Drosophila stem cell niches, are some very well-investigated stem cell niches. The niche components in the majority of systems include: (i) stem cells and progenitors (supply autocrine and paracrine provisions among their own cell lines); (ii) adjoining stromal cells (mesenchymal cells that supply paracrine sensory information); (iii) the ECM (constituting integrins); and (iv) extrinsic sensory data (from the inside or outside of the tissue, including blood arteries, nerves, or lymphocytes). The generation of better models of these niches, including signals, sources of such signals, and direct interactions between stem cells and cellular and non-cellular niche elements, will be required to fully understand the underlying dynamics in maintaining modulation, internal balance, and responding to tissue injury. Certain stem cell niche areas, including epidermal and hematopoietic, are further along than others, particularly the intestinal niche cells (Walker et al., 2009). More recent research is looking into how niche disruptions can lead to stem cell disorders, such as those seen in aging or cancer progression, considering the importance of the niche microenvironment in stem cell function (Pajonk & Vlashi, 2013; Rezza, Sennett, & Rendl, 2014).

6.1 ECM topography and material

Extracellular molecules and elements, namely, collagen, glycoproteins, etc., make up the ECM, a three-dimensional (3D) structure that offers molecular and structural stability to neighboring cells (Theocharis, Skandalis, Gialeli, & Karamanos, 2016). Every cell is in junction with the ECM. It is essential for the organization of tissues, differentiation of cells, and function. The ECM consists of proteoglycans (PG) and fibrous tissues, particularly their interlocking mesh. PGs occupy the vast extrinsic interspatial areas in tissues, such as hydrated gels. PGs have a broad spectrum of uses because of their inherent buffering, hydrating, bonding, and coercing characteristics (Frantz, Stewart, & Weaver, 2010). ECM components include:

6.1.1. Carbohydrate polymers like glycosaminoglycans (GAGs) constitute proteoglycans as they are linked to extracellular matrix proteins. The net negative charge of proteoglycans tends to captivate the sodium ions, which are positively charged (Na+), that further draw in water particles through diffusion, sustaining the ECM and some local cells by hydrating them. Proteoglycans may even assist the ECM in managing and store growth factors. Heparan sulfate, chondroitin sulfate, keratan sulfate, dermatan sulfate, and hyaluronic acid are examples of GAGs found in the extracellular matrix. Hyaluronic acid is the only GAG biosynthesized at the cell membrane rather than the Golgi apparatus and exists in a protein-free condition. Keratan sulfate is a structurally unique GAG comprising of D-galactose and not hexuronic acid in its disaccharide unit.

6.1.2. Collagen is perhaps the most widespread fibrous protein throughout the interspatial ECM, representing roughly 30% of the overall protein concentration of a multicellular organism. Collagens are the principal structural components of the ECM and render durability and cell adhesion regulation and aid in chemotaxis, migration, and driving tissue growth.

6.1.3. Elastin, linked to collagen, is another critical ECM fiber. Tissues that are stretched repeatedly benefit from the recoil provided by elastin fibers. Notably, the close connection of elastin with collagen fibrils limits elastin strain.

6.1.4. Fibronectin (FN), a third fibrous protein, plays an essential part in controlling the interspatial ECM's architecture and regulating cell adhesion and activity. Cell mechanical factors may stretch FN numerous times throughout its resting extent.

6.1.5. Laminins are mainly present in practically all organisms' basal laminae. Instead of generating collagen-like fibers, laminins in the basal lamina produce web-like arrangements which can withstand tensile stresses. They additionally help cells stick together. Many ECM constituents, including collagens, are bound by laminins.

The ECM's physical, biochemical, and biomechanical features are dictated by its constituents' morphological, metabolic, and operational complexity. Rigidity, permeability, topography, and low solubility are all physical qualities that might impact anchorage-related physiological activities such as division and migration of cells and polarity of the tissues (Gattazzo, Urciuolo, & Bonaldo, 2014). The ECM produces the physiochemical and mechanical characteristics of each tissue, such as compressive and flexural strengths and elasticity, and also facilitates coverage through a stabilizing action that upholds extracellular balance in the body and fluid retention through these physicochemical and biological characteristics (Ahmed, & ffrench-Constant, C., 2016). It seems to be a rapidly evolving framework that is also routinely reconstructed, whether enzymatically or nonenzymatically, and its individual constituents undergo various post-translational alterations.

6.2 Scaffold biomaterial, natural and synthetic

ECM is a natural scaffold and biomaterial meticulously examined and mimicked to create effective scaffolds for tissue regeneration or drug therapy. For years, the ECM has been investigated as a biomaterial in many aspects and formulations. ECM-like biomaterials are free of lipids, decellularized protein-based composites comprising purified polypeptide isolates from formerly functioning tissues or organs (Hinderer, Layland, & Schenke-Layland, 2016). Natural ECM gels, including type I collagen or matrigel, or fibrin, are composed of proteins that self-construct *in vitro* to form superior nanofibrous frameworks that closely resemble several *in vivo* conditions. They were the earliest materials that indicate that rigidity had an effect on cell fate. According to studies in collagen and fibrin gels, increased matrix crosslinking that regulates matrix rigidity affects integrin signaling and actomyosin-regulated cell stress, which are crucial characteristics in cancer development. Standard cell division and multiplication are also influenced by ECM protein gels (Hinderer et al., 2016). The majorly prevalent structural protein in soft tissues is type I collagen. It establishes a 3D setting for cells, allowing them to develop and impact their structure and activity. To generate a basement membrane-type scaffold, Laminin 111, a significant epithelial ECM constituent, has been employed as a part of Matrigel. It has also been utilized in electrospinning to build fibrous meshes having variable pore diameters and fiber sizes on a charged surface. *In vitro*, the designed meshes were able to keep adipose-derived stem cells attached and viable and stimulate them to produce neurite-type structures (Xing, Lee, Luo, & Kyriakides, 2020).

Regenerative approaches, often including stem cell-associated therapeutics and bioengineering techniques, are currently being investigated to rebuild, modify, repair, or maintain injured cells and tissues. For the synthesis of bio-fabricated replacement tissues, the construction of suitable 3D scaffolds, in conjunction with meticulous cell type choice, is critical. These frameworks may frequently be made of biodegradable or non-biodegradable polymers (Asti & Gioglio, 2018). To obtain superior mechanical stability and scaffold durability, synthetic biomaterials like polyurethane, polyethylene glycol (PEGs), or poly-L-lactide are utilized. Natural biomaterials such as alginate, chitosan, etc., have also been utilized (Goyal et al., 2017).

7. Culture systems and morphogens to study lung stem cell niche

7.1 2D and 3D cell culture system

Cell culture is a versatile tool used for basic scientific and translation research works. The technique has been used majorly for studying and investigating physiology and diseases. The practice of isolating epithelial cells from human sources has enabled the researchers to study various morphology and interactions between cells and between cells and stimuli (Rachamalla, Mukherjee, & Paul, 2021; Vertrees, Jordan, Solley, & Goodwin, 2009). The method dates back to the 20th century when it was used to study the growth of tissues and their maturation, virology, genes playing a role in disease development, and hybrid cell lines for the production of biopharmaceuticals (Segeritz & Vallier, 2017). Harrison et al. in 1907, made use of cell culture techniques in order to study the morphology and physiology of nerve fibers, after which it became a much-used method with significant improvements and advancements. Alex Carrel, in around 1912, began culturing of small heart tissue of an 18-day old chick embryo. This eventually made him the first scientist to culture mammalian cells in the laboratory (Chaicharoenaudomrung, Kunhorm, & Noisa, 2019; Segeritz & Vallier, 2017).

Several *in vitro* culture systems are used, cell, organ, *ex vivo*, and primary explant culture. Additionally, depending on the source, there are various cell culturing types like "primary cell culture" and "cell lines" (Rachamalla et al., 2021). The cells for primary cell culture are isolated directly from the concerned organism by tissue disaggregation possessing a limited timespan, whereas the cell lines are considered immortal, thereby being less problematic (Chaicharoenaudomrung et al., 2019; Vertrees et al., 2009). There are two main types of cell culture model systems employed, namely, 2-D and 3-D cell cultures, the former consists of cells arranged in a monolayer on flat and rigid substrates, whereas the latter contains spheroid structures with cells forming various layers (Edmondson, Broglie, Adcock, & Yang, 2014; Kapałczyńska et al., 2016). With the advent of cell culture techniques, the 2D monolayers were majorly used compared to 3D cultures to research cell-based assays as it is an easy, convenient, and inexpensive method. However, on the other hand, it also possesses certain limitations, one being its failure to mimic the cells living in the original *in vivo* microenvironment, thereby having a high probability of misleading results of the proposed hypotheses. It is supported by clear evidence through an experiment conducted on cancer cells grown in a 2D monolayer that resulted in drug response failure by nearly 95% of the cells, therefore representing a poor model for drug development (Chaicharoenaudomrung et al., 2019; Rachamalla et al., 2021). The 3D cell culture is known to be developed as a step ahead of the 2D culture system and has given unique cell-based assays which have proved to be more relevant in both morphological and behavioral functions when compared to the *in vivo* cells. Over the last couple of years, various *in vitro* platforms have been made to apply 3D cultures in different fields like drug development, cancer, and stem cell research, tissue engineering, and other experimental assays (Chaicharoenaudomrung et al., 2019; Rachamalla et al., 2021; Segeritz & Vallier, 2017).

7.2 Organoids

Organoids are 3D structures containing various types of cells to study embryonic and tissue development in artificial conditions (Mahapatra,

Lee, & Paul, 2022). 3D organoid technology is considered superior to conventional 2D monolayer cultures (Hofer & Lutolf, 2021; Mukherjee, Sinha, Maibam, Bisht, & Paul, 2022). They have been used in drug screening, modeling cancer, and hereditary diseases, transcriptome profiling, and cell or gene therapy (Ho, Pek, & Soh, 2018). Certain features define organoids and their relevance in molecular biology and tissue culture—it is synthesized from stem cells and/or primary tissues, it is cultured in pre-defined artificial conditions, mimic healthy and diseased tissues, show various complexities of modeled tissues, and can be given or shared as a whole culture or *via* progenitor cells (Lehmann et al., 2019). One of the most outstanding features of organoids is that they can be easily manipulated genetically, enabling one to study congenital disabilities, attack pathogens, and the efficacy of the drug(s) in living beings (Hofer & Lutolf, 2021; Takebe & Wells, 2019). Organoids have even found a great application in tracking the path of viral infection caused by SARS-COV 2 in COVID-19 (Sanyal & Paul, 2021). It helps identify infected organs and reactions involving pathological damage, thereby facilitating and accelerating the process of vaccine development and taking necessary control measures. Upon analysis of nearly 25 different cell lines by Chu et al., it was seen that the lungs and the small intestine are the most vulnerable organs to infection (Yu, 2021). In the current times, various organoids have developed so far in the liver, kidney, brain, lung, and intestine. Liver organoids are less time-consuming and highly effective in tracking infection by pathogens (including SARS-COV-2) in both liver and bile ducts that majorly result in liver damage. Kidney organoid cultures are helpful in assessing ACE 2 expression levels prior to infections. Organoid cultures of the brain enable one to study essential features of the developing brain and several neurological disorders (Chhibber et al., 2020). Pulmonary organoids act as a functional research model for various studies of respiratory diseases caused due to respiratory syncytial, avian influenza, and parainfluenza viruses. Culturing gut organoids facilitates research on gut inflammation and ulcers (Wang et al., 2017; Yu, 2021).

7.3 Morphogens and peptide

Morphogens are defined as synthesized locally signaling molecules that diffuse and act over long distances to induce cellular responses and control growth and pattern throughout a tissue region. The types of signaling molecules received by the cells then differentiate the progenitor cells into discrete cell types. These signaling molecules tend to create a concentration gradient in space for the differentiation of cells in the body of multicellular organisms (Sagner & Briscoe, 2017; Tabata & Takei, 2004). During the development of a multicellular organism, morphogens are synthesized and secreted from specific progenitor cells that are received by specific cell receptors for the activation of signaling cascades required for the growth and development of tissues and organs (Fung & Tsukamoto, 2015). The relevance and formation of concentration gradient by the cells solely regulate the desired patterning of the tissue. This usually occurs at various stages-synthesis, diffusion and degradation rate, and interaction with molecules present extracellularly that can act as agonists or antagonists for signal transduction across the receiving cells (Dekanty & Milán, 2011). Lipid–modified morphogens that are anchored to cell membranes, like Hh (Hedgehogs) and Wg (Wingless), are responsible for the patterning of tissues and the progression of diseases and disorders. It was first identified in *Drosophila*, where they activated paracrine signaling to develop the larval imaginal discs, resulting in the formation of adult body parts. In similar ways, Hh in vertebrates. Sonic hedgehogs (Shh) and homologs of Wg (Wnt) play a role in the development of limbs, neural tube, endocrine system, and craniofacial features. In grown-up adults, these kinds of morphogens facilitate the maintenance of stem cells, repair tissues, and keep steady metabolism (Takashima, Paul, Aghajanian, Younossi-Hartenstein, & Hartenstein, 2013). The Hh and Wg proteins are associated with various carrier molecules like exosomes, exovesicles, cytonemes, and lipoproteins. Exosomes are essential in immunological processes, paracrine signaling, and tumor-stroma crosstalk (Chauhan, Mudaliar, Basu, Aich, & Paul, 2022; Mukherjee, Bisht, Dutta, & Paul, 2022). Exovesicles were mainly involved in the release of Shh. Cytonemes assist in transportation Wg, Hh, Notch, fibroblast growth factor, epidermal growth factor, and Decapentaplegic. The secretion of Hh *via* lipoproteins was primarily reported during the overexpression of Hh in fat bodies (Parchure, Vyas, & Mayor, 2018).

The latest research has highlighted the importance and mechanism of morphogen–related targets in liver fibrosis. The liver has an attractive capacity among organs to regenerate itself after injury. This regeneration involves proliferation and differentiation of cells, followed by reactivation of morphogens to serve liver injuries. In an adult, normal individual, morphogen release is at extremely low levels. However, in the diseased states, morphogens like Necdin, DLK1, Wnt, Shh, and Notch are highly expressed and released by the hepatic stellate cells of the liver (Fung & Tsukamoto, 2015).

The use of proteins for therapeutic intervention has proved to be of great potential. The cause of many human diseases is mainly due to the improper functioning of proteins. However, its therapy is considered to be safer than gene-based therapies. The use of protein therapy is cost-effective, takes reasonable time for development and approval, and can easily be patent-protective. Cell-penetrating peptides (CPPs), a peptide 5–30 amino acids long, positively charged when recorded for physiological pH due to the presence of lysine or arginine, have been used to deliver various molecules like large active proteins for entering into cells by the process of endocytosis. Similarly, Cyclic CPP has promoted the cytosolic delivery of several proteins and RNAs (Kurrikoff, Vunk, & Langel, 2020; Patel et al., 2019; Ruseska & Zimmer, 2020). Apart from these, one of the peptides used for nasal drug delivery is Cyclodextrins. It is used as an adsorption-enhancing molecule to increase the bioavailability of proteins and peptide drugs in the intranasal region of the respiratory system (Meredith, Salameh, & Banks, 2015; Zidovetzki & Levitan, 2007).

8. Artificial scaffold and lung stem/progenitor cell niche

The lung has a critical architecture needed for its function. Researchers have shifted their interest from 2D cultures (i.e., 2D biology) into 3D space to replicate lung tissue physiology *in vitro*. Human lung models produced from tissue-specific stem/progenitor cells and hPSCs have been created to mimic structural and cellular aspects of the human lung.

8.1 Scaffold-based engineered lung organoid

Stem/progenitor cells grown in a biomaterial scaffold are used to grow lung organoids. Wilkinson et al., bioengineered lung organoids mimicked the structure of actual lungs. They used the self-assemble feature of adult lung stem cells-coated alginate beads in a bioreactor to create three-dimensional patterns similar to that of the human lung (Wilkinson et al., 2017, 2018). The beads provided a niche for the lab-grown lung-like tissue and may be used to investigate lung disorders that are difficult to examine using standard 2D models (Fig. 2) (Wilkinson et al., 2017). The role of Wnt/β-catenin in the development of BPD was studied using a 3D human organoid BPD model. This study demonstrates that aberrant Wnt/β-catenin signaling in BPD is suggestive of the role of a diseased niche (Sucre et al., 2017). Dye et al. used a highly microporous poly(lactide-*co*-glycolide) (PLG) scaffold coated with matrigel to grow lung organoids and cultured them both *in vitro* and *in vivo*. The PLG scaffold and matrigel provided a structural niche for the growth of the stem cells (Fig. 3) (Dye et al., 2016). Miller et al. generated human pluripotent stem cells (hPSCs) derived ventral-anterior

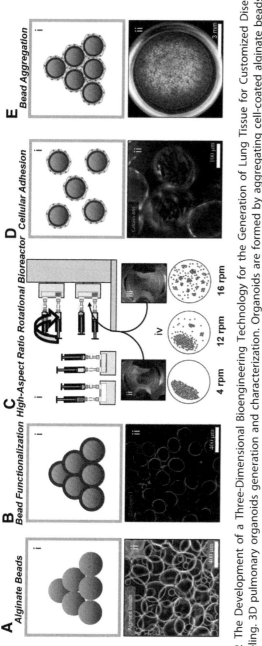

Fig. 2 The Development of a Three-Dimensional Bioengineering Technology for the Generation of Lung Tissue for Customized Disease Modeling. 3D pulmonary organoids generation and characterization. Organoids are formed by aggregating cell-coated alginate beads in a gently rotating HARV bioreactor or a 96-well plate configuration. Alginate bead graphic (Ai). (Aii) Micrograph of alginate beads in white light (scale bar = 400 μm). (Bi) Illustration of collagen I–coated alginate beads. (Bii) Immunofluorescence of collagen I, demonstrating a conformal coating of collagen I on the surface of the bead. Inset, high-resolution confocal z-stack of a single collagen-coated bead (scale bar = 400 μm). (Ci) HARV bioreactor loading and operation. In a 4-mL bioreactor vessel, 1 mL of sedimented, functionalized alginate beads was added. The tube was seeded with two million fetal lung fibroblasts. Rotation began once the vessel was fastened to the rotary base. (Cii) Time-lapse picture of beads moving in unison at 4 rpm in a 4-mL HARV bioreactor. (Ciii) Image of beads rotating independently at 16 rpm in a 4-mL HARV bioreactor. (Civ) A graphical representation of the bead flow patterns at various rpm levels. (Di) Illustration of beads coated with fetal lung fibroblasts following incubation in the HARV bioreactor. (Dii) Fluorescence micrograph of calcein-AM (viability dye) demonstrating equally coated fetal lung fibroblasts. (Ei) Illustration of aggregated beads covered with fetal lung fibroblasts. (Eii) After 3 days of culture, a typical mesenchymal 3D lung organoid was formed in the 96-well bioreactor (size bar = 3 mm). 3D, Three-dimensional; HARV, High-aspect-ratio vessel. *Figure reproduced with permission from Wilkinson, D. C., Alva-Ornelas, J. A., Sucre, J. M. S., Vijayaraj, P., Durra, A., Richardson, W., Jonas, S. J., Paul, M. K., Karumbayaram, S., Dunn, B., & Gomperts, B. N. (2017). Development of a three-dimensional bioengineering technology to generate lung tissue for personalized disease modeling. Stem Cells Translational Medicine, 6(2), 622–633. https://doi.org/10.5966/sctm.2016-0192.*

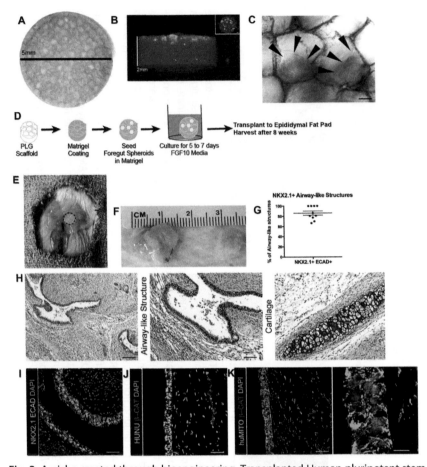

Fig. 3 A niche created through bioengineering. Transplanted Human pluripotent stem cell (hPSC)-derived human lung organoids (HLOs)-scaffold constructs grew, engrafted, and developed airway-like structures. (A) A 5 mm diameter PLG scaffold with a honeycomb-patterned architecture. (B) The majority of Di-O labeled 1-day HLOs (green) remained at the scaffold's surface, with only a few organoids ascending to the scaffold's center. The inset shows an aerial view of the scaffold, which is dotted with 1 day HLOs (green). (C) 1-day HLOs adhered to the scaffold's pores. The scale bar represents 100 mm. (D) PLG scaffolds were seeded with 1-day HLOs and cultured *in vitro* for 5–7 days in FGF10-supplemented media. After 8 weeks, the HLO-loaded scaffolds were transplanted into the mouse epididymal fat pad and harvested. (E) The HLO scaffold (dotted line) was inserted into the epididymal fat pad of a mouse. (F) Transplanted HLOs (tHLOs) ranged in length from 0.5 to 1.5 cm. (G) The average percentage of airway-like structures that were NKX2.1 + ECAD+ was 86.19 +/− 4.14% ($N=10$; error bars represent SEM). (H) H and E staining of tHLOs revealed airway-like structures (right two panels, low and high magnification) and cartilage pockets (left panel). At low magnification, the scale bar corresponds to 200 mm; at high magnification, it corresponds to

foregut spheroids. They further differentiated them into human lung organoids (Miller et al., 2019). Collagens (primarily Type I collagen), glycoproteins (such as laminin and FN), proteoglycans, and other substances, such as elastin and hyaluronic acid, make up most of the ECM and may be a critical component for recapitulating lung stem cell artificial niche (Dye et al., 2020). Hegab et al., used a 3D organoid model and different growth factors and coculture to imitating the niche of lung epithelial stem cells. Lung epithelial cells when cocultured with fibroblasts and different niche-secreted factors, yielded three different types of organoid. The type A were morphologically big, round, and with a central lumen; type B is big, irregular, with varying lumen thickness; while type C was small, rounded and without a lumen. and elucidating that the molecular mechanisms underlying their *in vitro* behavior (Fig. 4). This method established a way to bioengineer the lung stem cell niche and studied them *in vitro* (Hegab et al., 2015).

8.2 Bioprinting

Organoids that are cell-based 3D organ-mimetic vital in disease modeling and drug discovery can now be printed using 3D printing technology. Since 1984, when Charles Hull invented stereolithography (SLA) and patented his claims, the technique has evolved from basic shapes printed with resin and plastic to biological scaffolds printed with cells. 3D printing technique allows for the rapid fabrication of objects that closely mirror the architecture of biological tissue (Bisht, Hope, Mukherjee, & Paul, 2021). Traditionally, cells were seeded on prepared scaffold material and then implanted into the patient. Bioprinting is a promising new technology that allows all tissue components, including cells and their matrices, to be printed and patterned in three dimensions (Bisht, Hope, & Paul, 2019). Bioprinting improves cell deposition accuracy within prefabricated scaffolds, integrates

---

100 mm. (I) ECAD-outlined airway-like structures (white) expressed the lung marker NKX2.1 (green). The scale bar represents 50 mm. (J–K) The epithelium (b-CAT, red) and mesenchyme both expressed the human nuclear marker HUNU (J, green), as well as the human mitochondrial marker huMITO (K, green). The scale bars indicate 50 mm in J–K and 10 mm in the high magnification image in K. hPSC, Human pluripotent stem cell; HLO, human lung organoid; PLG, poly(lactide-*co*-glycolide); tHLO, Transplanted HLO. *Figure reproduced with permission from Dye, B. R., Dedhia, P. H., Miller, A. J., Nagy, M. S., White, E. S., Shea, L. D., & Spence, J. R. (2016). A bioengineered niche promotes in vivo engraftment and maturation of pluripotent stem cell derived human lung organoids. eLife, 5. https://doi.org/10.7554/eLife.19732.*

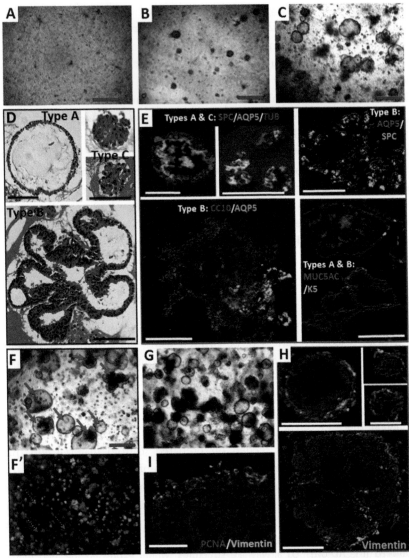

Fig. 4 Establishment of 3D-organoid culture conditions for studying niche interactions of lung stem cells *in vitro*. Lung epithelial cells (50,000) were cultured alone (A) or in coculture with 10:1 (B) or 1:1 (C) mouse lung fibroblasts for 2 weeks. (D) Colonies growing in 1:1 coculture wells exhibited three distinct morphologies, designated A, B, and C. (E) Immunofluorescent staining of the three colony types revealed their differentiation spectrum for the various lung epithelial cell markers. (F and F') Lung epithelial cells from SP-C/GFP transgenic mice were cocultured with wild-type lung fibroblasts; only Type C colonies expressed GFP, whereas all type A and B colonies expressed GFP (red arrows in (F) indicate some type A and type B colonies. The dotted arrows in (F')

numerous cell types and matrix materials in a scaffold, and engineers scaffold shape and anatomical precision. Furthermore, bioprinting can provide patient-specific characteristics in a scaffold while regulating pore size, shape, and connection. The following section describes the history and methods of 3D printing.

The primary bioprinting technologies are droplet-based, extrusion-based, and laser-assisted bioprinting (Fig. 5). The popular printers employ inkjet or droplet technology. These printers create biological material on a *Z*-adjustable electrically controlled elevator platform, much like 2D printers. High pressure is used to split a bioink solution into droplets and deposit it in a continuous or drop-on-demand mode. Drop-on-demand

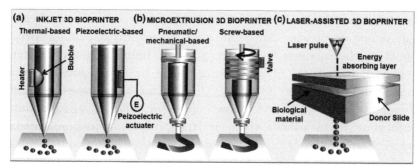

Fig. 5 Cartoon illustration of a schematic of 3D bioprinting technologies (A) Inkjet/droplet-based 3D bioprinting; (B) extrusion-based bioprinting; and (C) Laser-based bioprinting. There is a comparison of important aspects of bioprinting. *Figure reproduced with permission from Bisht, B., Hope, A., Mukherjee, A., & Paul, M. K. (2021). Advances in the fabrication of scaffold and 3D printing of biomimetic bone graft. Annals of Biomedical Engineering, 49(4), 1128–1150. https:/doi.org/10.1007/s10439-021-02752-9.*

indicate the location of the same colonies that are GFP positive). (G) Coculture of lung epithelial cells and tracheal fibroblasts. When lung fibroblasts were cocultured, this favored the growth of more type A colonies than type B and C colonies. (H) Mesenchymal cell marker staining of the three organoid types with vimentin demonstrated that fibroblasts tightly wrap themselves around the three distinct colony types. (I) Vimentin and PCNA staining revealed that epithelial cells adjacent to the fibroblast wrapping proliferate, but more central cells do not. Scale bars: 500 μm for (A–C, F, and G); 50 μm for the remainder. *Figure reproduced with permission from Hegab, A. E., Arai, D., Gao, J., Kuroda, A., Yasuda, H., Ishii, M., Naoki, K., Soejima, K., & Betsuyaku, T. (2015). Mimicking the niche of lung epithelial stem cells and characterization of several effectors of their in vitro behavior. Stem Cell Research, 15(1), 109–121. https:/doi.org/10.1016/j.scr.2015.05.005.*

bioprinting is preferred for precise stacking and mixing. By dispensing ink, inkjet bioprinters are classed as thermal or piezoelectric. Ink droplets are forced out of the extrusion nozzle by a small vapor bubble that swiftly grows and bursts. Droplets of bioink are generated when the voltage pulses distort the piezoelectric element, altering the volume of the fluid chamber and creating a pressure wave that overcomes surface tension at the nozzle, thereby forcing out a droplet. Extrusion-based printing uses an auxiliary pressure system with a 3-axis robot to layer live cells and scaffold materials in continuous lines creating a base layer, followed by layer addition over the base, to form a 3D structure. Cells in polymer or hydrogel are extruded in lines by pneumatic, piston, or screw-driven force on a build plate. Pneumatic techniques are sluggish and best for viscous molten hydrogels. Moreover, piston-driven dispensers provide greater control of the hydrogel flow *via* the nozzle. For greater viscosity hydrogels, screw-based methods are helpful but may generate significant pressure dips at the nose. Hence, the dispensing systems should be chosen carefully depending on the biopolymer and cell type employed. Laser-based 3D bioprinting works by laser-induced forward transfer and direct lasers to deposit droplets for bioprinting. Originally designed to move metal, and is now utilized for biologics. The bioprinter is a focusing system, a donor, a biological transport support (ribbon) material, and a collector substrate facing the fabric. A single pulse is used to focus on the quartz ribbon-absorption layer resulting in the creation of volatile bubbles that facilitate the material transfer. As low-power UV lasers cause relatively shallow penetration, direct effects on the cells by the laser are avoided. This method is also known as laser-guided direct writing (LGDW). A laser beam can accurately etch droplets on a base plate (Ren et al., 2021). Other bioink and cell transfer methods to transfer include laser-guided direct writing (LGDW), matrix-assisted pulsed-laser evaporation-direct write (MAPLE-DW), and laser-induced forward transfer (LIFT).

3D bioprinting employs 3D printing and 3D printing-like methods to create biological tissues and artificial organs. Printing 3D organoids can help rapidly develop disease models as they may print increasingly complex structures. In order to understand microenvironmental control, different cell lines embedded in designed hydrogels can be printed simultaneously or individually, creating 3D-mini organoids. The use of 3D organoids and tumoroids can speed up drug development and personalized therapy. Though the creation of 3D printed lung organoids is still a matter of investigation, several attempts have been made to create 3D printed lung organoids. Taniguchi et al. used rat chondrocytes, endothelial cells, and MSCs to generate spheroids. The spheroids were robotically articulated and 3Dprinted using a Regenova bio-3D printer to build a scaffold-free tubular artificial rat tracheal construct. The bioprinted construct underwent spheroid fusion and maturation in a bioreactor for 28 days, while care was taken to avoid lumen compaction. The isogenic scaffold-free tracheal structure was evaluated for mechanical testing and grafting in rats, and further research can evaluate vascularization and implant aspects (Taniguchi et al., 2018). Machino et al. improved on the earlier strategy by using human cartilage, fibroblast, umbilical vein endothelial, and bone marrow MSCs to produce seven types of spheroids followed by 3D printing scaffold-free trachea-like cartilaginous constructs. Growing the 3D printed construct in a bioreactor for 28 days facilitates cell rearrangement and self-organization, boosting mechanical strength and tissue function. Cellular analysis of the transplanted constructs in rats suggests cellular aggregation, proliferation, and nascent vascular development (Machino et al., 2019). Though this process may be effective for larger airways but generating smaller airways is a significant challenge. Alternative assistance mechanisms should be researched to solve this issue. Recently, Grigoryan et al. used stereolithography to construct and test soft hydrogel-based topologies mimicking a multivascularized alveolar model. They also generated a computational model to study airway inflation and oxygenation efficiencies (Grigoryan et al., 2019). This model could mimic alveolar vasculature topologies, but 3D printing distal lung organoids remain challenging. Other printing techniques, such as the freeform reversible embedding of suspended hydrogels (FRESH) approach for printing low viscosity bioinks, require support baths of sacrificial materials, such as a slurry of gelatin microparticles (Galliger, Vogt, & Panoskaltsis-Mortari, 2019; Ren et al., 2021).

9. Conclusion and future direction

Recent research has helped identify diverse lung stem/progenitor cell types and ways to promote endogenous regeneration, but lung transplantation is the only treatment for most chronic and acute lung disorders. Donor lungs are in limited supply. On top of that, the 5-year survival rate following lung transplantation is just 50% (Tsuchiya, Doi, Obata, Hatachi, & Nagayasu, 2020). A viable alternative option is lung tissue regeneration. Regenerative therapy has the potential to heal a wide variety of previously incurable ailments and, therefore, the need for further understanding of the stem/progenitor cell niche. Humanized models are the need of the day and hence the need to elucidate biomaterial and fabricated systems that can mimic human lung stem/progenitor cells niche. Designing microfluidic systems, microplates, and plastic wares that can help create an appropriate stem cell niche microenvironment to recreate cell–cell and cell–matrix interactions can be vital in recreating artificial niches. The use of nanomaterials, advanced materials, and nanotechnology approaches can help pattern specific factors, scaffold material, and cells in a niche-compatible geometry to fast-track high-throughput screening. 3D bioprinting has excellent advantages in efficiently layering scaffolds, extracellular matrix, proteins, and cells to create artificial niches.

Advances in 3D culture and organoid systems have opened up the option of using multiple cell type cocultures to investigate the different molecular mediators in stem niches. The geometry, peptide presentation, stiffness, molecular cues, and scaffolds may be altered and titrated to better understand stem/progenitor cell potency, self-renewal, and lineage specification. To decipher small molecule niche modulators, high-throughput techniques are also needed. Another exciting aspect of tackling lung diseases is stem cell transplantation and regenerative therapy. Understanding the stem cell niches can provide promising inputs for the survival and expansion of transplanted edited/normal stem cells in various lung disorders. The lung, a complex organ system, is challenging to manage in 3D culture systems, especially concerning respiratory functions. Another aspect is modeling diseased niches to understand their role in disease pathogenesis is also critical. The advances in lung artificial stem cell niche generation can be extended to model other stem cell niches. In conclusion, with the current advances in material science, nanotechnology, 3D-culture system, stem cell biology, single-cell omics, and 3D printing, the future for creating artificial niches is promising. It will accelerate drug screening and generate avenues for regenerative therapies targeting lung diseases. Consideration of advancements in stem cell biology, nanotechnology, and material chemistry may aid in the recapitulation of the lung stem cell niche and aid in the development of successful medicines in the near future.

Acknowledgments

B.B. acknowledges Dr. Jay M. Lee for providing research support. M.K.P. acknowledges S. Dubinett and B. Gomperts from UCLA for providing constant support and mentoring.

Competing interests

The authors declare no competing interests.

References

Ahmed, M., & ffrench-Constant, C. (2016). Extracellular matrix regulation of stem cell behavior. *Current Stem Cell Reports, 2*(3), 197–206. https://doi.org/10.1007/s40778-016-0056-2.

Akram, K., Patel, N., Spiteri, M., & Forsyth, N. (2016). Lung regeneration: Endogenous and exogenous stem cell mediated therapeutic approaches. *International Journal of Molecular Sciences, 17*(1). https://doi.org/10.3390/ijms17010128.

Aros, C. J., Paul, M. K., Pantoja, C. J., Bisht, B., Meneses, L. K., Vijayaraj, P., et al. (2020). High-throughput drug screening identifies a potent Wnt inhibitor that promotes airway basal stem cell homeostasis. *Cell Reports, 30*(7), 2055–2064. e2055. https://doi.org/10.1016/j.celrep.2020.01.059.

Aros, C. J., Vijayaraj, P., Pantoja, C. J., Bisht, B., Meneses, L. K., Sandlin, J. M., et al. (2020). Distinct spatiotemporally dynamic Wnt-secreting niches regulate proximal airway regeneration and aging. *Cell Stem Cell, 27*(3), 413–429.e414. https://doi.org/10.1016/j.stem.2020.06.019.

Asti, A., & Gioglio, L. (2018). Natural and synthetic biodegradable polymers: Different scaffolds for cell expansion and tissue formation. *The International Journal of Artificial Organs, 37*(3), 187–205. https://doi.org/10.5301/ijao.5000307.

Barkauskas, C. E., Cronce, M. J., Rackley, C. R., Bowie, E. J., Keene, D. R., Stripp, B. R., et al. (2013). Type 2 alveolar cells are stem cells in adult lung. *Journal of Clinical Investigation, 123*(7), 3025–3036. https://doi.org/10.1172/jci68782.

Barnes, P. J. (2016). Inflammatory mechanisms in patients with chronic obstructive pulmonary disease. *Journal of Allergy and Clinical Immunology, 138*(1), 16–27. https://doi.org/10.1016/j.jaci.2016.05.011.

Beronja, S., & Fuchs, E. (2011). A breath of Fresh air in lung regeneration. *Cell, 147*(3), 485–487. https://doi.org/10.1016/j.cell.2011.10.008.

Bertoncello, I. (2016). Properties of adult lung stem and progenitor cells. *Journal of Cellular Physiology, 231*(12), 2582–2589. https://doi.org/10.1002/jcp.25404.

Birbrair, A., & Frenette, P. S. (2016). Niche heterogeneity in the bone marrow. *Annals of the New York Academy of Sciences, 1370*(1), 82–96. https://doi.org/10.1111/nyas.13016.

Bisht, B., Hope, A., Mukherjee, A., & Paul, M. K. (2021). Advances in the fabrication of scaffold and 3D printing of biomimetic bone graft. *Annals of Biomedical Engineering, 49*(4), 1128–1150. https://doi.org/10.1007/s10439-021-02752-9.

Bisht, B., Hope, A., & Paul, M. K. (2019). From papyrus leaves to bioprinting and virtual reality: History and innovation in anatomy. *Anatomy & Cell Biology, 52*(3). https://doi.org/10.5115/acb.18.213.

Brandt, J. P., & Mandiga, P. (2021). *Histology, alveolar cells*. StatPearls. (https://www.ncbi.nlm.nih.gov/pubmed/32491474).

Busch, S. M., Lorenzana, Z., & Ryan, A. L. (2021). Implications for extracellular matrix interactions with human lung basal stem cells in lung development, disease, and airway modeling. *Frontiers in Pharmacology, 12*. https://doi.org/10.3389/fphar.2021.645858.

Carraro, G., Mulay, A., Yao, C., Mizuno, T., Konda, B., Petrov, M., et al. (2020). Single-cell reconstruction of human basal cell diversity in Normal and idiopathic pulmonary fibrosis lungs. *American Journal of Respiratory and Critical Care Medicine, 202*(11), 1540–1550. https://doi.org/10.1164/rccm.201904-0792OC.

Chaicharoenaudomrung, N., Kunhorm, P., & Noisa, P. (2019). Three-dimensional cell culture systems as an in vitro platform for cancer and stem cell modeling. *World Journal of Stem Cells, 11*(12), 1065–1083. https://doi.org/10.4252/wjsc.v11.i12.1065.

Chatterjee, A., Paul, S., Bisht, B., Bhattacharya, S., Sivasubramaniam, S., & Paul, M. K. (2022). Advances in targeting the WNT/β-catenin signaling pathway in cancer. *Drug Discovery Today, 27*(1), 82–101. https://doi.org/10.1016/j.drudis.2021.07.007.

Chauhan, D. S., Mudaliar, P., Basu, S., Aich, J., & Paul, M. K. (2022). Tumor-derived exosome and immune modulation. In *Extracellular vesicles - role in diseases pathogenesis and therapy [working title]*. https://doi.org/10.5772/intechopen.103718.

Chen, L., Acciani, T., Le Cras, T., Lutzko, C., & Perl, A.-K. T. (2012). Dynamic regulation of platelet-derived growth factor receptor α expression in alveolar fibroblasts during Realveolarization. *American Journal of Respiratory Cell and Molecular Biology, 47*(4), 517–527. https://doi.org/10.1165/rcmb.2012-0030OC.

Chhibber, T., Bagchi, S., Lahooti, B., Verma, A., Al-Ahmad, A., Paul, M. K., et al. (2020). CNS organoids: An innovative tool for neurological disease modeling and drug neurotoxicity screening. *Drug Discovery Today, 25*(2), 456–465. https://doi.org/10.1016/j.drudis.2019.11.010.

Choi, J., Jang, Y. J., Dabrowska, C., Iich, E., Evans, K. V., Hall, H., et al. (2021). Release of notch activity coordinated by IL-1β signalling confers differentiation plasticity of airway progenitors via Fosl2 during alveolar regeneration. *Nature Cell Biology, 23*(9), 953–966. https://doi.org/10.1038/s41556-021-00742-6.

Choi, J., Park, J.-E., Tsagkogeorga, G., Yanagita, M., Koo, B.-K., Han, N., et al. (2020). Inflammatory signals induce AT2 cell-derived damage-associated transient progenitors that mediate alveolar regeneration. *Cell Stem Cell, 27*(3), 366–382. e367. https://doi.org/10.1016/j.stem.2020.06.020.

Cortesi, E., & Ventura, J. J. (2019). Lgr6: From stemness to cancer progression. *Journal of Lung Health and Diseases, 3*(1), 12–15. https://www.ncbi.nlm.nih.gov/pubmed/31236545.

De Santis, M. M., Bölükbas, D. A., Lindstedt, S., & Wagner, D. E. (2018). How to build a lung: Latest advances and emerging themes in lung bioengineering. *European Respiratory Journal, 52*(1). https://doi.org/10.1183/13993003.01355-2016.

Dekanty, A., & Milán, M. (2011). The interplay between morphogens and tissue growth. *EMBO Reports, 12*(10), 1003–1010. https://doi.org/10.1038/embor.2011.172.

Desai, T. J., Brownfield, D. G., & Krasnow, M. A. (2014). Alveolar progenitor and stem cells in lung development, renewal and cancer. *Nature, 507*(7491), 190–194. https://doi.org/10.1038/nature12930.

Donne, M. L., Lechner, A. J., & Rock, J. R. (2015). Evidence for lung epithelial stem cell niches. *BMC Developmental Biology, 15*(1). https://doi.org/10.1186/s12861-015-0082-9.

Dye, B. R., Dedhia, P. H., Miller, A. J., Nagy, M. S., White, E. S., Shea, L. D., et al. (2016). A bioengineered niche promotes in vivo engraftment and maturation of pluripotent stem cell derived human lung organoids. *eLife, 5*. https://doi.org/10.7554/eLife.19732.

Dye, B. R., Youngblood, R. L., Oakes, R. S., Kasputis, T., Clough, D. W., Spence, J. R., et al. (2020). Human lung organoids develop into adult airway-like structures directed by physico-chemical biomaterial properties. *Biomaterials, 234*. https://doi.org/10.1016/j.biomaterials.2020.119757.

Edmondson, R., Broglie, J. J., Adcock, A. F., & Yang, L. (2014). Three-dimensional cell culture systems and their applications in drug discovery and cell-based biosensors. *Assay and Drug Development Technologies, 12*(4), 207–218. https://doi.org/10.1089/adt.2014.573.

El Agha, E., Herold, S., Alam, D. A., Quantius, J., MacKenzie, B., Carraro, G., et al. (2014). Fgf10-positive cells represent a progenitor cell population during lung development and postnatally. *Development, 141*(2), 296–306. https://doi.org/10.1242/dev.099747.

Frantz, C., Stewart, K. M., & Weaver, V. M. (2010). The extracellular matrix at a glance. *Journal of Cell Science, 123*(24), 4195–4200. https://doi.org/10.1242/jcs.023820.

Fung, E., & Tsukamoto, H. (2015). Morphogen-related therapeutic targets for liver fibrosis. *Clinics and Research in Hepatology and Gastroenterology, 39*, S69–S74. https://doi.org/10.1016/j.clinre.2015.05.017.

Galliger, Z., Vogt, C. D., & Panoskaltsis-Mortari, A. (2019). 3D bioprinting for lungs and hollow organs. *Translational Research*, *211*, 19–34. https://doi.org/10.1016/j.trsl.2019.05.001.

Gattazzo, F., Urciuolo, A., & Bonaldo, P. (2014). Extracellular matrix: A dynamic microenvironment for stem cell niche. *Biochimica et Biophysica Acta (BBA) - General Subjects*, *1840*(8), 2506–2519. https://doi.org/10.1016/j.bbagen.2014.01.010.

Ghaedi, M., Le, A. V., Hatachi, G., Beloiartsev, A., Rocco, K., Sivarapatna, A., et al. (2017). Bioengineered lungs generated from human iPSC s-derived epithelial cells on native extracellular matrix. *Journal of Tissue Engineering and Regenerative Medicine*, *12*(3). https://doi.org/10.1002/term.2589.

Gomperts, B. N. (2014). Induction of multiciliated cells from induced pluripotent stem cells. *Proceedings of the National Academy of Sciences*, *111*(17), 6120–6121. https://doi.org/10.1073/pnas.1404414111.

Gomperts, B. N., & Strieter, R. M. (2007). Stem cells and chronic lung disease. *Annual Review of Medicine*, *58*(1), 285–298. https://doi.org/10.1146/annurev.med.58.081905.134954.

Goyal, R., Vega, M. E., Pastino, A. K., Singh, S., Guvendiren, M., Kohn, J., et al. (2017). Development of hybrid scaffolds with natural extracellular matrix deposited within synthetic polymeric fibers. *Journal of Biomedical Materials Research Part A*, *105*(8), 2162–2170. https://doi.org/10.1002/jbm.a.36078.

Grigoryan, B., Paulsen, S. J., Corbett, D. C., Sazer, D. W., Fortin, C. L., Zaita, A. J., et al. (2019). Multivascular networks and functional intravascular topologies within biocompatible hydrogels. *Science*, *364*(6439), 458–464. https://doi.org/10.1126/science.aav9750.

Guha, A., Vasconcelos, M., Cai, Y., Yoneda, M., Hinds, A., Qian, J., et al. (2012). Neuroepithelial body microenvironment is a niche for a distinct subset of Clara-like precursors in the developing airways. *Proceedings of the National Academy of Sciences*, *109*(31), 12592–12597. https://doi.org/10.1073/pnas.1204710109.

Gupte, V. V., Ramasamy, S. K., Reddy, R., Lee, J., Weinreb, P. H., Violette, S. M., et al. (2009). Overexpression of fibroblast growth Factor-10 during both inflammatory and fibrotic phases attenuates bleomycin-induced pulmonary fibrosis in mice. *American Journal of Respiratory and Critical Care Medicine*, *180*(5), 424–436. https://doi.org/10.1164/rccm.200811-1794OC.

Han, F., Wang, J., Ding, L., Hu, Y., Li, W., Yuan, Z., et al. (2020). Tissue engineering and regenerative medicine: Achievements, future, and sustainability in Asia. *Frontiers in Bioengineering and Biotechnology*, *8*. https://doi.org/10.3389/fbioe.2020.00083.

Harding, R., & Hooper, S. B. (1996). Regulation of lung expansion and lung growth before birth. *Journal of Applied Physiology*, *81*(1), 209–224. https://doi.org/10.1152/jappl.1996.81.1.209.

Hegab, A. E., Arai, D., Gao, J., Kuroda, A., Yasuda, H., Ishii, M., et al. (2015). Mimicking the niche of lung epithelial stem cells and characterization of several effectors of their in vitro behavior. *Stem Cell Research*, *15*(1), 109–121. https://doi.org/10.1016/j.scr.2015.05.005.

Hegab, A. E., Ha, V. L., Bisht, B., Darmawan, D. O., Ooi, A. T., Zhang, K. X., et al. (2014). Aldehyde dehydrogenase activity enriches for proximal airway basal stem cells and promotes their proliferation. *Stem Cells and Development*, *23*(6), 664–675. https://doi.org/10.1089/scd.2013.0295.

Hegab, A. E., Ha, V. L., Darmawan, D. O., Gilbert, J. L., Ooi, A. T., Attiga, Y. S., et al. (2012). Isolation and in vitro characterization of basal and submucosal gland duct stem/progenitor cells from human proximal airways. *Stem Cells Translational Medicine*, *1*(10), 719–724. https://doi.org/10.5966/sctm.2012-0056.

Herriges, M., & Morrisey, E. E. (2014). Lung development: Orchestrating the generation and regeneration of a complex organ. *Development, 141*(3), 502–513. https://doi.org/10.1242/dev.098186.

Hilgendorff, A., Parai, K., Ertsey, R., Navarro, E., Jain, N., Carandang, F., et al. (2015). Lung matrix and vascular remodeling in mechanically ventilated elastin haploinsufficient newborn mice. *American Journal of Physiology. Lung Cellular and Molecular Physiology, 308*(5), L464–L478. https://doi.org/10.1152/ajplung.00278.2014.

Hinderer, S., Layland, S. L., & Schenke-Layland, K. (2016). ECM and ECM-like materials — Biomaterials for applications in regenerative medicine and cancer therapy. *Advanced Drug Delivery Reviews, 97*, 260–269. https://doi.org/10.1016/j.addr.2015.11.019.

Ho, B., Pek, N., & Soh, B.-S. (2018). Disease modeling using 3D organoids derived from human induced pluripotent stem cells. *International Journal of Molecular Sciences, 19*(4). https://doi.org/10.3390/ijms19040936.

Hofer, M., & Lutolf, M. P. (2021). Engineering organoids. *Nature Reviews Materials, 6*(5), 402–420. https://doi.org/10.1038/s41578-021-00279-y.

Hoffman, A. M., Shifren, A., Mazan, M. R., Gruntman, A. M., Lascola, K. M., Nolen-Walston, R. D., et al. (2010). Matrix modulation of compensatory lung regrowth and progenitor cell proliferation in mice. *American Journal of Physiology. Lung Cellular and Molecular Physiology, 298*(2), L158–L168. https://doi.org/10.1152/ajplung.90594.2008.

Ishizawa, K., Kubo, H., Yamada, M., Kobayashi, S., Numasaki, M., Ueda, S., et al. (2004). Bone marrow-derived cells contribute to lung regeneration after elastase-induced pulmonary emphysema. *FEBS Letters, 556*(1–3), 249–252. https://doi.org/10.1016/s0014-5793(03)01399-1.

Jain, R., Barkauskas, C. E., Takeda, N., Bowie, E. J., Aghajanian, H., Wang, Q., et al. (2015). Plasticity of Hopx+ type I alveolar cells to regenerate type II cells in the lung. *Nature. Communications, 6*(1). https://doi.org/10.1038/ncomms7727.

Jhala, D., & Vasita, R. (2015). A review on extracellular matrix mimicking strategies for an artificial stem cell niche. *Polymer Reviews, 55*(4), 561–595. https://doi.org/10.1080/15583724.2015.1040552.

Kapałczyńska, M., Kolenda, T., Przybyła, W., Zajączkowska, M., Teresiak, A., Filas, V., et al. (2016). 2D and 3D cell cultures – A comparison of different types of cancer cell cultures. *Archives of Medical Science*. https://doi.org/10.5114/aoms.2016.63743.

Kathiriya, J. J., & Peng, T. (2021). An inflammatory switch for stem cell plasticity. *Nature Cell Biology, 23*(9), 928–929. https://doi.org/10.1038/s41556-021-00752-4.

Katsura, H., Kobayashi, Y., Tata, P. R., & Hogan, B. L. M. (2019). IL-1 and TNFα contribute to the inflammatory niche to enhance alveolar regeneration. *Stem Cell Reports, 12*(4), 657–666. https://doi.org/10.1016/j.stemcr.2019.02.013.

Kim, C. F. B., Jackson, E. L., Woolfenden, A. E., Lawrence, S., Babar, I., Vogel, S., et al. (2005). Identification of bronchioalveolar stem cells in Normal lung and lung cancer. *Cell, 121*(6), 823–835. https://doi.org/10.1016/j.cell.2005.03.032.

Kotton, D. N., & Morrisey, E. E. (2014). Lung regeneration: Mechanisms, applications and emerging stem cell populations. *Nature Medicine, 20*(8), 822–832. https://doi.org/10.1038/nm.3642.

Krause, A., Conboy, M. J., & Conboy, I. M. (2017). Cellular senescence and stem cell niche. In *Biology and engineering of stem cell niches* (pp. 185–192). https://doi.org/10.1016/b978-0-12-802734-9.00012-3.

Kumar, P. A., Hu, Y., Yamamoto, Y., Hoe, N. B., Wei, T. S., Mu, D., et al. (2011). Distal airway stem cells yield alveoli in vitro and during lung regeneration following H1N1 influenza infection. *Cell, 147*(3), 525–538. https://doi.org/10.1016/j.cell.2011.10.001.

Kurrikoff, K., Vunk, B., & Langel, Ü. (2020). Status update in the use of cell-penetrating peptides for the delivery of macromolecular therapeutics. *Expert Opinion on Biological Therapy, 21*(3), 361–370. https://doi.org/10.1080/14712598.2021.1823368.

Labaki, W. W., & Han, M. K. (2020). Chronic respiratory diseases: A global view. *The Lancet Respiratory Medicine*, *8*(6), 531–533. https://doi.org/10.1016/s2213-2600(20)30157-0.

Lechner, A. J., Driver, I. H., Lee, J., Conroy, C. M., Nagle, A., Locksley, R. M., et al. (2017). Recruited monocytes and type 2 immunity promote lung regeneration following pneumonectomy. *Cell Stem Cell*, *21*(1), 120–134.e127. https://doi.org/10.1016/j.stem.2017.03.024.

Lee, J.-H., Bhang, D. H., Beede, A., Huang, T. L., Stripp, B. R., Bloch, K. D., et al. (2014). Lung stem cell differentiation in mice directed by endothelial cells via a BMP4-NFATc1-Thrombospondin-1 Axis. *Cell*, *156*(3), 440–455. https://doi.org/10.1016/j.cell.2013.12.039.

Lee, J.-H., & Rawlins, E. L. (2018). Developmental mechanisms and adult stem cells for therapeutic lung regeneration. *Developmental Biology*, *433*(2), 166–176. https://doi.org/10.1016/j.ydbio.2017.09.016.

Lehmann, R., Lee, C. M., Shugart, E. C., Benedetti, M., Charo, R. A., Gartner, Z., et al. (2019). Human organoids: A new dimension in cell biology. *Molecular Biology of the Cell*, *30*(10), 1129–1137. https://doi.org/10.1091/mbc.E19-03-0135.

Li, F., He, J., Wei, J., Cho, W. C., & Liu, X. (2015). Diversity of epithelial stem cell types in adult lung. *Stem Cells International*, *2015*, 1–11. https://doi.org/10.1155/2015/728307.

Li, J., Wang, Z., Chu, Q., Jiang, K., Li, J., & Tang, N. (2018). The strength of mechanical forces determines the differentiation of alveolar epithelial cells. *Developmental Cell*, *44*(3), 297–312. e295. https://doi.org/10.1016/j.devcel.2018.01.008.

Little, B. P., Junn, J. C., Zheng, K. S., Sanchez, F. W., Henry, T. S., Veeraraghavan, S., et al. (2020). Diffuse idiopathic pulmonary neuroendocrine cell hyperplasia: Imaging and clinical features of a frequently delayed diagnosis. *American Journal of Roentgenology*, *215*(6), 1312–1320. https://doi.org/10.2214/ajr.19.22628.

Liu, K., Tang, M., Liu, Q., Han, X., Jin, H., Zhu, H., et al. (2020). Bi-directional differentiation of single bronchioalveolar stem cells during lung repair. *Cell. Discovery*, *6*(1). https://doi.org/10.1038/s41421-019-0132-8.

Logan, C. Y., & Desai, T. J. (2015). Keeping it together: Pulmonary alveoli are maintained by a hierarchy of cellular programs. *BioEssays*, *37*(9), 1028–1037. https://doi.org/10.1002/bies.201500031.

Longmire, T. A., Ikonomou, L., Hawkins, F., Christodoulou, C., Cao, Y., Jean, J. C., et al. (2012). Efficient derivation of purified lung and thyroid progenitors from embryonic stem cells. *Cell Stem Cell*, *10*(4), 398–411. https://doi.org/10.1016/j.stem.2012.01.019.

Machino, R., Matsumoto, K., Taniguchi, D., Tsuchiya, T., Takeoka, Y., Taura, Y., et al. (2019). Replacement of rat tracheas by layered, trachea-like, scaffold-free structures of human cells using a bio-3D printing system. *Advanced Healthcare Materials*, *8*(7). https://doi.org/10.1002/adhm.201800983.

Mahapatra, C., Lee, R., & Paul, M. K. (2022). Emerging role and promise of nanomaterials in organoid research. *Drug Discovery Today*, *27*(3), 890–899. https://doi.org/10.1016/j.drudis.2021.11.007.

McQualter, J. L., Yuen, K., Williams, B., & Bertoncello, I. (2010). Evidence of an epithelial stem/progenitor cell hierarchy in the adult mouse lung. *Proceedings of the National Academy of Sciences*, *107*(4), 1414–1419. https://doi.org/10.1073/pnas.0909207107.

Melo-Narváez, M. C., Stegmayr, J., Wagner, D. E., & Lehmann, M. (2020). Lung regeneration: Implications of the diseased niche and ageing. *European Respiratory Review*, *29*(157). https://doi.org/10.1183/16000617.0222-2020.

Meredith, M. E., Salameh, T. S., & Banks, W. A. (2015). Intranasal delivery of proteins and peptides in the treatment of neurodegenerative diseases. *The AAPS Journal*, *17*(4), 780–787. https://doi.org/10.1208/s12248-015-9719-7.

Miller, A. J., Dye, B. R., Ferrer-Torres, D., Hill, D. R., Overeem, A. W., Shea, L. D., et al. (2019). Generation of lung organoids from human pluripotent stem cells in vitro. *Nature Protocols*, *14*(2), 518–540. https://doi.org/10.1038/s41596-018-0104-8.

Montoro, D. T., Haber, A. L., Biton, M., Vinarsky, V., Lin, B., Birket, S. E., et al. (2018). A revised airway epithelial hierarchy includes CFTR-expressing ionocytes. *Nature*, *560*(7718), 319–324. https://doi.org/10.1038/s41586-018-0393-7.

Mori, M., Mahoney, J. E., Stupnikov, M. R., Paez-Cortez, J. R., Szymaniak, A. D., Varelas, X., et al. (2015). Notch3-jagged signaling controls the pool of undifferentiated airway progenitors. *Development*, *142*(2), 258–267. https://doi.org/10.1242/dev. 116855.

Morrisey, E. E., & Hogan, B. L. M. (2010). Preparing for the first breath: Genetic and cellular mechanisms in lung development. *Developmental Cell*, *18*(1), 8–23. https://doi.org/10. 1016/j.devcel.2009.12.010.

Mou, H., Zhao, R., Sherwood, R., Ahfeldt, T., Lapey, A., Wain, J., et al. (2012). Generation of multipotent lung and airway progenitors from mouse ESCs and patient-specific cystic fibrosis iPSCs. *Cell Stem Cell*, *10*(4), 385–397. https://doi.org/10.1016/j.stem. 2012.01.018.

Mukherjee, A., Bisht, B., Dutta, S., & Paul, M. K. (2022). Current advances in the use of exosomes, liposomes, and bioengineered hybrid nanovesicles in cancer detection and therapy. *Acta Pharmacologica Sinica*. https://doi.org/10.1038/s41401-022-00902-w.

Mukherjee, A., Sinha, A., Maibam, M., Bisht, B., & Paul, M. K. (2022). Organoids and commercialization. In *Organoids* (pp. 1–17). IntechOpen. https://doi.org/https://doi.org/10. 5772/intechopen.104706 (IntechOpen).

Nabhan, A. N., Brownfield, D. G., Harbury, P. B., Krasnow, M. A., & Desai, T. J. (2018). Single-cell Wnt signaling niches maintain stemness of alveolar type 2 cells. *Science*, *359*(6380), 1118–1123. https://doi.org/10.1126/science.aam6603.

Nikolić, M. Z., Sun, D., & Rawlins, E. L. (2018). Human lung development: Recent progress and new challenges. *Development*, *145*(16). https://doi.org/10.1242/dev.163485.

Noguchi, M., Furukawa, K. T., & Morimoto, M. (2020). Pulmonary neuroendocrine cells: Physiology, tissue homeostasis and disease. *Disease Models & Mechanisms*, *13*(12). https://doi.org/10.1242/dmm.046920.

Oeztuerk-Winder, F., Guinot, A., Ochalek, A., & Ventura, J.-J. (2012). Regulation of human lung alveolar multipotent cells by a novel p38α MAPK/miR-17-92 axis. *The EMBO Journal*, *31*(16), 3431–3441. https://doi.org/10.1038/emboj.2012.192.

Olajuyin, A. M., Zhang, X., & Ji, H.-L. (2019). Alveolar type 2 progenitor cells for lung injury repair. *Cell death. Discovery*, *5*(1). https://doi.org/10.1038/s41420-019-0147-9.

Ouadah, Y., Rojas, E. R., Riordan, D. P., Capostagno, S., Kuo, C. S., & Krasnow, M. A. (2019). Rare pulmonary neuroendocrine cells are stem cells regulated by Rb, p53, and notch. *Cell*, *179*(2), 403–416. e423. https://doi.org/10.1016/j.cell.2019.09.010.

Pajonk, F., & Vlashi, E. (2013). Characterization of the stem cell niche and its importance in radiobiological response. *Seminars in Radiation Oncology*, *23*(4), 237–241. https://doi.org/ 10.1016/j.semradonc.2013.05.007.

Panganiban, R. A. M., & Day, R. M. (2010). Hepatocyte growth factor in lung repair and pulmonary fibrosis. *Acta Pharmacologica Sinica*, *32*(1), 12–20. https://doi.org/10.1038/ aps.2010.90.

Parchure, A., Vyas, N., & Mayor, S. (2018). Wnt and hedgehog: Secretion of lipid-modified morphogens. *Trends in Cell Biology*, *28*(2), 157–170. https://doi.org/10.1016/j.tcb.2017. 10.003.

Pardo-Saganta, A., Law, B. M., Tata, P. R., Villoria, J., Saez, B., Mou, H., et al. (2015). Injury induces direct lineage segregation of functionally distinct airway basal stem/progenitor cell subpopulations. *Cell Stem Cell*, *16*(2), 184–197. https://doi.org/10.1016/j. stem.2015.01.002.

Pardo-Saganta, A., Tata, P. R., Law, B. M., Saez, B., Chow, R. D.-W., Prabhu, M., et al. (2015). Parent stem cells can serve as niches for their daughter cells. *Nature, 523*(7562), 597–601. https://doi.org/10.1038/nature14553.

Patel, S. G., Sayers, E. J., He, L., Narayan, R., Williams, T. L., Mills, E. M., et al. (2019). Cell-penetrating peptide sequence and modification dependent uptake and subcellular distribution of green florescent protein in different cell lines. *Scientific Reports, 9*(1). https://doi.org/10.1038/s41598-019-42456-8.

Paul, M. K., Bisht, B., Darmawan, D. O., Chiou, R., Ha, V. L., Wallace, W. D., et al. (2014). Dynamic changes in intracellular ROS levels regulate airway basal stem cell homeostasis through Nrf2-dependent notch signaling. *Cell Stem Cell, 15*(2), 199–214. https://doi.org/10.1016/j.stem.2014.05.009.

Peng, T., Frank, D. B., Kadzik, R. S., Morley, M. P., Rathi, K. S., Wang, T., et al. (2015). Hedgehog actively maintains adult lung quiescence and regulates repair and regeneration. *Nature, 526*(7574), 578–582. https://doi.org/10.1038/nature14984.

Pennings, S., Liu, K. J., & Qian, H. (2018). The stem cell niche: Interactions between stem cells and their environment. *Stem Cells International, 2018*, 1–3. https://doi.org/10.1155/2018/4879379.

Rabata, A., Fedr, R., Soucek, K., Hampl, A., & Koledova, Z. (2020). 3D cell culture models demonstrate a role for FGF and WNT signaling in regulation of lung epithelial cell fate and morphogenesis. *Frontiers in Cell and Development Biology, 8*. https://doi.org/10.3389/fcell.2020.00574.

Rachamalla, H., Mukherjee, A., & Paul, M. K. (2021). Nanotechnology application and intellectual property right prospects of mammalian cell culture. In *Cell culture [working title]*. https://doi.org/10.5772/intechopen.99146.

Ratajczak, M. Z., Ratajczak, J., Suszynska, M., Miller, D. M., Kucia, M., & Shin, D.-M. (2017). A novel view of the adult stem cell compartment from the perspective of a quiescent population of very small embryonic-like stem cells. *Circulation Research, 120*(1), 166–178. https://doi.org/10.1161/circresaha.116.309362.

Ren, Y., Yang, X., Ma, Z., Sun, X., Zhang, Y., Li, W., et al. (2021). Developments and opportunities for 3D bioprinted organoids. *International Journal of Bioprinting, 7*(3). https://doi.org/10.18063/ijb.v7i3.364.

Reynolds, S. D., Giangreco, A., Power, J. H. T., & Stripp, B. R. (2000). Neuroepithelial bodies of pulmonary airways serve as a reservoir of progenitor cells capable of epithelial regeneration. *The American Journal of Pathology, 156*(1), 269–278. https://doi.org/10.1016/s0002-9440(10)64727-x.

Rezza, A., Sennett, R., & Rendl, M. (2014). Adult Stem Cell Niches. In *Stem cells in development and disease* (pp. 333–372). https://doi.org/https://doi.org/10.1016/b978-0-12-416022-4.00012-3.

Rock, J. R., Onaitis, M. W., Rawlins, E. L., Lu, Y., Clark, C. P., Xue, Y., et al. (2009). Basal cells as stem cells of the mouse trachea and human airway epithelium. *Proceedings of the National Academy of Sciences, 106*(31), 12771–12775. https://doi.org/10.1073/pnas.0906850106.

Rock, J. R., Randell, S. H., & Hogan, B. L. M. (2010). Airway basal stem cells: A perspective on their roles in epithelial homeostasis and remodeling. *Disease Models & Mechanisms, 3*(9–10), 545–556. https://doi.org/10.1242/dmm.006031.

Ruiz, E. J., Oeztuerk-Winder, F., & Ventura, J.-J. (2014). A paracrine network regulates the cross-talk between human lung stem cells and the stroma. *Nature. Communications, 5*(1). https://doi.org/10.1038/ncomms4175.

Ruseska, I., & Zimmer, A. (2020). Internalization mechanisms of cell-penetrating peptides. *Beilstein Journal of Nanotechnology, 11*, 101–123. https://doi.org/10.3762/bjnano.11.10.

Sagner, A., & Briscoe, J. (2017). Morphogen interpretation: Concentration, time, competence, and signaling dynamics. *Wiley interdisciplinary reviews. Developmental Biology, 6*(4). https://doi.org/10.1002/wdev.271.

Sala, F. G., Del Moral, P.-M., Tiozzo, C., Alam, D. A., Warburton, D., Grikscheit, T., et al. (2011). FGF10 controls the patterning of the tracheal cartilage rings via shh. *Development, 138*(2), 273–282. https://doi.org/10.1242/dev.051680.

Sanyal, R., & Paul, M. K. (2021). Organoid technology and the COVID pandemic. In *Origin and impact of COVID-19 pandemic originating from SARS-CoV-2 infection across the globe [working title]*. https://doi.org/https://doi.org/10.5772/intechopen.98542.

Sauler, M., McDonough, J. E., Adams, T. S., Kothapalli, N., Barnthaler, T., Werder, R. B., et al. (2022). Characterization of the COPD alveolar niche using single-cell RNA sequencing. *Nature. Communications, 13*(1). https://doi.org/10.1038/s41467-022-28062-9.

Schiller, H. B., Fernandez, I. E., Burgstaller, G., Schaab, C., Scheltema, R. A., Schwarzmayr, T., et al. (2015). Time- and compartment-resolved proteome profiling of the extracellular niche in lung injury and repair. *Molecular Systems Biology, 11*(7). https://doi.org/10.15252/msb.20156123.

Schittny, J. C. (2017). Development of the lung. *Cell and Tissue Research, 367*(3), 427–444. https://doi.org/10.1007/s00441-016-2545-0.

Segeritz, C.-P., & Vallier, L. (2017). Cell Culture. In *Basic science methods for clinical researchers* (pp. 151–172). https://doi.org/10.1016/b978-0-12-803077-6.00009-6.

Shah, U., Rahulan, V., Kumar, P., Dutta, P., & Attawar, S. (2021). Donor lung management: Changing perspectives. *Lung India, 38*(5). https://doi.org/10.4103/lungindia.lungindia_476_20.

Smirnova, N. F., Schamberger, A. C., Nayakanti, S., Hatz, R., Behr, J., & Eickelberg, O. (2016). Detection and quantification of epithelial progenitor cell populations in human healthy and IPF lungs. *Respiratory Research, 17*(1). https://doi.org/10.1186/s12931-016-0404-x.

Song, H., Yao, E., Lin, C., Gacayan, R., Chen, M.-H., & Chuang, P.-T. (2012). Functional characterization of pulmonary neuroendocrine cells in lung development, injury, and tumorigenesis. *Proceedings of the National Academy of Sciences, 109*(43), 17531–17536. https://doi.org/10.1073/pnas.1207238109.

Stabler, C. T., & Morrisey, E. E. (2016). Developmental pathways in lung regeneration. *Cell and Tissue Research, 367*(3), 677–685. https://doi.org/10.1007/s00441-016-2537-0.

Stephenson, M., & Grayson, W. (2018). Recent advances in bioreactors for cell-based therapies. *F1000Research, 7*. https://doi.org/10.12688/f1000research.12533.1.

Sucre, J. M. S., Vijayaraj, P., Aros, C. J., Wilkinson, D., Paul, M., Dunn, B., et al. (2017). Posttranslational modification of β-catenin is associated with pathogenic fibroblastic changes in bronchopulmonary dysplasia. *American Journal of Physiology. Lung Cellular and Molecular Physiology, 312*(2), L186–L195. https://doi.org/10.1152/ajplung.00477.2016.

Sui, P., Wiesner, D. L., Xu, J., Zhang, Y., Lee, J., Van Dyken, S., et al. (2018). Pulmonary neuroendocrine cells amplify allergic asthma responses. *Science, 360*(6393). https://doi.org/10.1126/science.aan8546.

Surate Solaligue, D. E., Rodríguez-Castillo, J. A., Ahlbrecht, K., & Morty, R. E. (2017). Recent advances in our understanding of the mechanisms of late lung development and bronchopulmonary dysplasia. *American Journal of Physiology. Lung Cellular and Molecular Physiology, 313*(6), L1101–L1153. https://doi.org/10.1152/ajplung.00343.2017.

Tabata, T., & Takei, Y. (2004). Morphogens, their identification and regulation. *Development, 131*(4), 703–712. https://doi.org/10.1242/dev.01043.

Takashima, S., Paul, M., Aghajanian, P., Younossi-Hartenstein, A., & Hartenstein, V. (2013). Migration of drosophila intestinal stem cells across organ boundaries. *Development, 140*(9), 1903–1911. https://doi.org/10.1242/dev.082933.

Takebe, T., & Wells, J. M. (2019). Organoids by design. *Science, 364*(6444), 956–959. https://doi.org/10.1126/science.aaw7567.

Taniguchi, D., Matsumoto, K., Tsuchiya, T., Machino, R., Takeoka, Y., Elgalad, A., et al. (2018). Scaffold-free trachea regeneration by tissue engineering with bio-3D printing†. *Interactive Cardiovascular and Thoracic Surgery, 26*(5), 745–752. https://doi.org/10.1093/icvts/ivx444.

Tata, P. R., Mou, H., Pardo-Saganta, A., Zhao, R., Prabhu, M., Law, B. M., et al. (2013). Dedifferentiation of committed epithelial cells into stem cells in vivo. *Nature, 503*(7475), 218–223. https://doi.org/10.1038/nature12777.

Tata, P. R., & Rajagopal, J. (2017). Plasticity in the lung: Making and breaking cell identity. *Development, 144*(5), 755–766. https://doi.org/10.1242/dev.143784.

Theocharis, A. D., Skandalis, S. S., Gialeli, C., & Karamanos, N. K. (2016). Extracellular matrix structure. *Advanced Drug Delivery Reviews, 97*, 4–27. https://doi.org/10.1016/j.addr.2015.11.001.

Tiozzo, C., Langhe, S. D., Carraro, G., Alam, D. A., Nagy, A., Wigfall, C., et al. (2009). Fibroblast growth factor 10 plays a causative role in the tracheal cartilage defects in a mouse model of apert syndrome. *Pediatric Research, 66*(4), 386–390. https://doi.org/10.1203/PDR.0b013e3181b45580.

Tsao, P.-N., Matsuoka, C., Wei, S.-C., Sato, A., Sato, S., Hasegawa, K., et al. (2016). Epithelial notch signaling regulates lung alveolar morphogenesis and airway epithelial integrity. *Proceedings of the National Academy of Sciences, 113*(29), 8242–8247. https://doi.org/10.1073/pnas.1511236113.

Tsuchiya, T., Doi, R., Obata, T., Hatachi, G., & Nagayasu, T. (2020). Lung microvascular niche, repair, and engineering. *Frontiers in Bioengineering and Biotechnology, 8*. https://doi.org/10.3389/fbioe.2020.00105.

Van Raemdonck, D., Neyrinck, A., Verleden, G. M., Dupont, L., Coosemans, W., Decaluwe, H., et al. (2009). Lung donor selection and management. *Proceedings of the American Thoracic Society, 6*(1), 28–38. https://doi.org/10.1513/pats.200808-098GO.

Van Winkle, L. S., Buckpitt, A. R., Nishio, S. J., Isaac, J. M., & Plopper, C. G. (1995). Cellular response in naphthalene-induced Clara cell injury and bronchiolar epithelial repair in mice. *American Journal of Physiology. Lung Cellular and Molecular Physiology, 269*(6), L800–L818. https://doi.org/10.1152/ajplung.1995.269.6.L800.

Varghese, B., Ling, Z., & Ren, X. (2022). Reconstructing the pulmonary niche with stem cells: A lung story. *Stem Cell Research & Therapy, 13*(1). https://doi.org/10.1186/s13287-022-02830-2.

Vaughan, A. E., Brumwell, A. N., Xi, Y., Gotts, J. E., Brownfield, D. G., Treutlein, B., et al. (2014). Lineage-negative progenitors mobilize to regenerate lung epithelium after major injury. *Nature, 517*(7536), 621–625. https://doi.org/10.1038/nature14112.

Verheyden, J. M., & Sun, X. (2020). A transitional stem cell state in the lung. *Nature Cell Biology, 22*(9), 1025–1026. https://doi.org/10.1038/s41556-020-0561-5.

Vertrees, R. A., Jordan, J. M., Solley, T., & Goodwin, T. J. (2009). Tissue Culture Models. In *Basic concepts of molecular pathology* (pp. 159–182). https://doi.org/10.1007/978-0-387-89626-7_18.

Vijayaraj, P., Minasyan, A., Durra, A., Karumbayaram, S., Mehrabi, M., Aros, C. J., et al. (2019). Modeling progressive fibrosis with pluripotent stem cells identifies an anti-fibrotic small molecule. *Cell Reports, 29*(11), 3488–3505. e3489. https://doi.org/10.1016/j.celrep.2019.11.019.

Volckaert, T., & De Langhe, S. (2014). Lung epithelial stem cells and their niches: Fgf10 takes center stage. *Fibrogenesis & Tissue Repair, 7*(1). https://doi.org/10.1186/1755-1536-7-8.

Volckaert, T., Dill, E., Campbell, A., Tiozzo, C., Majka, S., Bellusci, S., et al. (2011). Parabronchial smooth muscle constitutes an airway epithelial stem cell niche in the mouse lung after injury. *Journal of Clinical Investigation*, *121*(11), 4409–4419. https://doi.org/10.1172/jci58097.

Voswinckel, R. (2004). Characterisation of post-pneumonectomy lung growth in adult mice. *European Respiratory Journal*, *24*(4), 524–532. https://doi.org/10.1183/09031936.04.10004904.

Walker, M. R., Patel, K. K., & Stappenbeck, T. S. (2009). The stem cell niche. *The Journal of Pathology*, *217*(2), 169–180. https://doi.org/10.1002/path.2474.

Wang, Z., Wang, S.-N., Xu, T.-Y., Miao, Z.-W., Su, D.-F., & Miao, C.-Y. (2017). Organoid technology for brain and therapeutics research. *CNS Neuroscience & Therapeutics*, *23*(10), 771–778. https://doi.org/10.1111/cns.12754.

Watson, J. K., Rulands, S., Wilkinson, A. C., Wuidart, A., Ousset, M., Van Keymeulen, A., et al. (2015). Clonal dynamics reveal two distinct populations of basal cells in slow-turnover airway epithelium. *Cell Reports*, *12*(1), 90–101. https://doi.org/10.1016/j.celrep.2015.06.011.

Whitsett, J. A. (2014). The molecular era of surfactant biology. *Neonatology*, *105*(4), 337–343. https://doi.org/10.1159/000360649.

Whitsett, J. A., & Weaver, T. E. (2002). Hydrophobic surfactant proteins in lung function and disease. *New England Journal of Medicine*, *347*(26), 2141–2148. https://doi.org/10.1056/NEJMra022387.

Wilkinson, D. C., Alva-Ornelas, J. A., Sucre, J. M. S., Vijayaraj, P., Durra, A., Richardson, W., et al. (2017). Development of a three-dimensional bioengineering technology to generate lung tissue for personalized disease modeling. *Stem Cells Translational Medicine*, *6*(2), 622–633. https://doi.org/10.5966/sctm.2016-0192.

Wilkinson, D. C., Mellody, M., Meneses, L. K., Hope, A. C., Dunn, B., & Gomperts, B. N. (2018). Development of a three-dimensional bioengineering technology to generate lung tissue for personalized disease modeling. *Current Protocols in Stem Cell Biology*, *46*(1). https://doi.org/10.1002/cpsc.56.

Woods, J. C., Schittny, J. C., Jobe, A., Whitsett, J., & Abman, S. (2016). Lung structure at preterm and term birth. In *Fetal and neonatal lung development* (pp. 126–140). https://doi.org/10.1017/cbo9781139680349.008.

Xing, H., Lee, H., Luo, L., & Kyriakides, T. R. (2020). Extracellular matrix-derived biomaterials in engineering cell function. *Biotechnology Advances*, *42*. https://doi.org/10.1016/j.biotechadv.2019.107421.

Xu, J., Yu, H., & Sun, X. (2020). Less is more: Rare pulmonary neuroendocrine cells function as critical sensors in lung. *Developmental Cell*, *55*(2), 123–132. https://doi.org/10.1016/j.devcel.2020.09.024.

Yao, E., Lin, C., Wu, Q., Zhang, K., Song, H., & Chuang, P.-T. (2018). Notch signaling controls Transdifferentiation of pulmonary neuroendocrine cells in response to lung injury. *Stem Cells*, *36*(3), 377–391. https://doi.org/10.1002/stem.2744.

Yin, H., Jing, B., Xu, D., Guo, W., Sun, B., Zhang, J., et al. (2022). Identification of active bronchioalveolar stem cells as the cell of origin in lung adenocarcinoma. *Cancer Research*, *82*(6), 1025–1037. https://doi.org/10.1158/0008-5472.Can-21-2445.

Yu, J. (2021). Organoids: A new model for SARS-CoV-2 translational research. *International Journal of Stem Cells*, *14*(2), 138–149. https://doi.org/10.15283/ijsc20169.

Yuan, T., Volckaert, T., Chanda, D., Thannickal, V. J., & De Langhe, S. P. (2018). Fgf10 signaling in lung development, homeostasis, disease, and repair after injury. *Frontiers in Genetics*, *9*. https://doi.org/10.3389/fgene.2018.00418.

Yuan, T., Volckaert, T., Redente, E. F., Hopkins, S., Klinkhammer, K., Wasnick, R., et al. (2019). FGF10-FGFR2B signaling generates basal cells and drives alveolar epithelial regeneration by bronchial epithelial stem cells after lung injury. *Stem Cell Reports*, *12*(5), 1041–1055. https://doi.org/10.1016/j.stemcr.2019.04.003.

Zakrzewski, W., Dobrzyński, M., Szymonowicz, M., & Rybak, Z. (2019). Stem cells: Past, present, and future. *Stem Cell Research & Therapy*, *10*(1). https://doi.org/10.1186/s13287-019-1165-5.

Zepp, J. A., Zacharias, W. J., Frank, D. B., Cavanaugh, C. A., Zhou, S., Morley, M. P., et al. (2017). Distinct mesenchymal lineages and niches promote epithelial self-renewal and Myofibrogenesis in the lung. *Cell*, *170*(6), 1134–1148. e1110 https://doi.org/10.1016/j.cell.2017.07.034.

Zhang, Y., Goss, A. M., Cohen, E. D., Kadzik, R., Lepore, J. J., Muthukumaraswamy, K., et al. (2008). A Gata6-Wnt pathway required for epithelial stem cell development and airway regeneration. *Nature Genetics*, *40*(7), 862–870. https://doi.org/10.1038/ng.157.

Zidovetzki, R., & Levitan, I. (2007). Use of cyclodextrins to manipulate plasma membrane cholesterol content: Evidence, misconceptions and control strategies. *Biochimica et Biophysica Acta (BBA) - Biomembranes*, *1768*(6), 1311–1324. https://doi.org/10.1016/j.bbamem.2007.03.026.

CHAPTER FOUR

Engineering mammary tissue microenvironments *in vitro*

Julien Clegg[a,b,†], Maria Koch[a,†], Akhilandeshwari Ravichandran[a,b], Dietmar W. Hutmacher[a,b,c,d,*], and Laura J. Bray[a,b]

[a]School of Mechanical, Medical and Process Engineering, Queensland University of Technology (QUT), Brisbane, QLD, Australia
[b]ARC Training Centre for Cell and Tissue Engineering Technologies, Queensland University of Technology (QUT), Brisbane, QLD, Australia
[c]School of Biomedical Sciences, Queensland University of Technology, Brisbane, QLD, Australia
[d]Max Planck Queensland Center for Materials Science of Extracellular Matrices, Brisbane, QLD, Australia
*Corresponding author: e-mail address: dietmar.hutmacher@qut.edu.au

Contents

1.	Anatomy of the normal mammary microenvironment	145
2.	Development, progression and clinical management of BC	148
	2.1 Development and staging	148
	2.2 Tumor microenvironment	150
	2.3 Current treatments for BC	153
3.	3D tumor modeling	156
	3.1 Three-dimensional (3D) models	156
	3.2 3D models for BC	160
4.	Current limitations and considerations of 3D models for BC	169
5.	Future prospective of 3D BC modeling systems	170
	References	172

1. Anatomy of the normal mammary microenvironment

The female breast is a highly dynamic and complex organ which encompasses anatomical structures such as mammary glands, apocrine glands, complex vessel networks and supported by adipocyte populations (Geddes, 2007). Development of the breast and the mammary microenvironment is a tightly orchestrated series of events that begin as early as the fourth week of gestation and continues throughout the trimesters

---

[†] These authors contributed equally.

Advances in Stem Cells and their Niches, Volume 6
ISSN 2468-5097
https://doi.org/10.1016/bs.asn.2022.02.001

145

(Medina, 1996). By the second trimester, mammary buds begin to form and penetrate the dermis underlying the developing breast, which is partly facilitated by the presence of mesenchymal cells. Through the later parts of the second trimester and throughout the final trimester, the mammary glands, lactiferous ducts and tubular vessels form within the dense fibroconnective tissue. During this period, further differentiation of surrounding tissue populations continue and increase of vascularity of the fibroconnective tissue occurs which is assisted by maternal and fetal hormones (Jolicoeur, 2005; Jolicoeur, Gaboury, & Oligny, 2003).

Further, extracellular matrix (ECM) molecules such as proteoglycans (PGs) play a large role in the development of the fetal breast. Syndecan, a transmembrane matrix receptor PG synergistically works with integrins and transmembrane glycoprotein dystroglycan to assemble signaling platforms that ultimately assist in dictating cell fate, cell behavior and enabling adherence of cells to the surrounding ECM (Barresi & Campbell, 2006; Morgan, Humphries, & Bass, 2007). Perlecan, a heparan-sulfate PG (HSPG), is derived from the basement membrane and functions to capture stromal fibroblast-growth factor 10 (FGF-10), where it is delivered to the epithelium and is considered a critical cell-ECM protein in mammary development (Patel et al., 2007).

In the anatomy of the adult breast, the whole ductal system is lined by two epithelial cell types. Located internally are the luminal epithelial cells which can differentiate into lactocytes, milk-secreting cells, during pregnancy. The luminal layer is supported by the outer layer of myoepithelial cells and then the basement membrane overlays all the epithelial tissue (Hassiotou & Geddes, 2013; Ramsay, Kent, Hartmann, & Hartmann, 2005). The surrounding stromal tissue of the mammary gland consists mainly of fibroblasts. It creates the cellular microenvironment of the glandular tissue, along with adipocytes and immune cells, such as local macrophages. Furthermore, surrounding capillary blood vessels are important to transport oxygen and nutrients to the cells, as well as remove tissue waste products. Capillaries are built by aligned endothelial cells, surrounded by a basement membrane. They are mechanically stabilized and supported by mesenchymal stem cells (MSCs), pericytes, and occasionally by smooth muscle fibers. The extracellular matrix is an important acellular component of the mammary tissue microenvironment, which not only gives structural support but also provides access to the numerous growth factors (GFs) and PGs that interact with the cell types in the breast to mediate and regulate their functions (Hassiotou & Geddes, 2013; Joshi, Di Grappa, & Khokha, 2012).

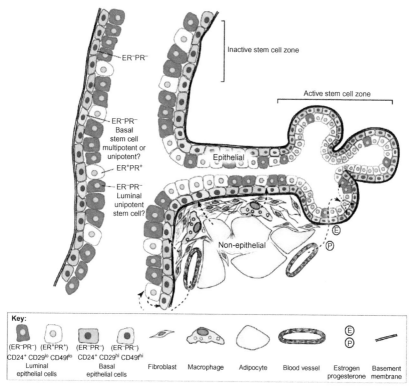

Fig. 1 The ductal system is made of two epithelial cell layers and is surrounded by stromal and vascular tissue made of non-epithelial cells (fibroblasts, adipocytes, macrophages). Cellular components of the mammary stem cell niche can generate signals to influence stem cell fate. Estrogen and progesterone act on luminal epithelial cells expressing estrogen (ER+) and progesterone receptors (PR+), which transfer the systemic hormone messages to ER-PR− mammary stem cells. Legend show illustrations of basic cells and factors in the anatomy of mammary lobules and the tissue microenvironment. *Figure reproduced with permission from Joshi, P. A., Di Grappa, M. A., & Khokha, R. (2012). Active allies: Hormones, stem cells and the niche in adult mammopoiesis.* Trends in Endocrinology & Metabolism, 23(6), 299–309.

Fig. 1 illustrates anatomical landmarks and structures of the adult breast which are also surrounding local and regionally placed lymph nodes and their respective networks (Joshi et al., 2012; PDQ Adult Treatment Editorial Board, 2020).

During pregnancy, further remodeling of the ECM occurs to accommodate for expected nourishment of the neonate. Increase in regional vascularization and mobilization of local and regional adipocytes prepares the organ for breast feeding, which this process is referred to as lactogenesis.

Adipocytes function as a local energy source for the intensive energy requirements of milk production (Macias & Hinck, 2012; Muschler & Streuli, 2010).

The breast and the mammary microenvironment are tightly controlled, consisting of both cellular and non-cellular components that remodel during different periods in a female's life and also sustain homeostatic conditions for proper function. Despite this, accumulative mutations and disruptions to cellular homeostasis can lead to aberrant behaviors of cells and the development of tumors and later breast cancer (BC). Due to the innate complexity of the breast, the subsequent BC that can form demonstrates innate heterogeneous and therefore remains difficult to understand and treat.

2. Development, progression and clinical management of BC

2.1 Development and staging

BC, like all cancers, develop due to mutations in the genome of single cells which can accumulate from factors such as age, race, lifestyle and environmental carcinogens. Additionally, there is a growing number of recognized subtypes of BC depending upon molecular signature and origins. The etiological importance of these growing number of subtypes further complicates how we understand the disease and how it is clinically treated, though may also give us a platform for specializing treatment regimes and developing novel drugs (Carey et al., 2006; Perou et al., 2000).

As mentioned, BC can be classified by a number of phenotypes and origins. Benign, malignant, invasive or non-invasive (*in situ*) are methods of describing gross behaviors of a tumor or cancer and the potential extent of its spreading. Lobular and ductal carcinomas, classified as adenocarcinomas, are the most common subtype of BC, though other BC cancers within the breast can occur such as stromal tumors, lymphomas or as a secondary site of metastasis from a different primary cancer (Sinn & Kreipe, 2013; Viale, 2012). BC can further be categorized upon the molecular markers that are present, which is referred to as molecular subtype, or subtyping based upon molecular signatures. There are three predominant molecular subtypes currently recognized which include hormonal receptor positive (HR+), human epidermal growth factor receptor 2 positive (HER2), or triple-negative (TNBC). Hormonal receptor positive BC refer to the upregulation and presence of estrogen or progesterone receptors which is similar with HER2 subtypes. A BC can possess all three receptor markers and those cancers that possess upregulated HER2 coincide with worser prognoses compared to only HR+. These two molecular subtypes can be targeted with targeted treatments to inhibit growth and kill cancerous cells, which is clinically beneficial. However, the third molecular subtype, TNBC, are BC that possess no upregulation or presence of these three molecular markers, which clinically translates to no targeted treatments and therefore, lower survival rates (Blows et al., 2010; Viale, 2012). Fig. 2 below shows the classification of BC subtypes according to immunohistochemistry (IHC) marker profiles as well as epithelial origin, summarizing the most common origins and molecular subtyping of BC.

The cells lining the lobules and ductal tubes account for 90% of all BC occurrences and are usually only detected following invasion of surrounding breast tissue. The breast is surrounded by a dense network of lymph vessels and nodes that also neighbors the anatomical sites of the sternum, neck and armpits. In many cases of metastatic BC, lymph nodes are the first site of colonization and invasion, thereby characterizing the cancer as *ex situ*,

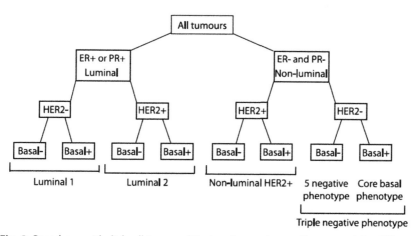

Fig. 2 Based on epithelial cell type and their cell receptor profile, up to eight different breast cancer types can be defined. Cancerous epithelial cells can be differentiated by the expression of the hormone receptors for estrogen (ER) and progesterone (PR) and can be subdivided by the additional expression of human epidermal GF receptor 2 (HER2). Further categorizations are based on basal marker characteristics. *Figure reproduced from Blows, F. M., Driver, K. E., Schmidt, M. K., Broeks, A., Van Leeuwen, F. E., Wesseling, J. et al. (2010). Subtyping of breast cancer by immunohistochemistry to investigate a relationship between subtype and short and long term survival: A collaborative analysis of data for 10,159 cases from 12 studies. PLoS Medicine, 7(5), e1000279.*

compared to the local non-invasive BC being referred to as *in situ*. Symptom onset usually occurs following this transition of *in-situ* to *ex situ* (Allred, 2010; Cancer Council, 2019).

The staging classification of the cancer will depend upon the developmental stage and the resulting migration or lack thereof (Chatterjee & McCaffrey, 2014). According to this figure, BC possesses a number of developmental stages, from being a benign epithelial tumor such as atypical hyperplasia to an *in situ* malignant carcinoma to an *ex situ* invasive cancer. These stages are used to better inform the course of treatment and therapies that might be required and include 0-IV. Stage 0 refers to pre-invasive cancer such as a ductal carcinoma *in situ* (DCIS) or lobular carcinoma *in situ* (LCIS), while stage 1A refers to early BC and 1B being advanced BC. Stages 2-onwards thereby, refers to advanced BCs of increasing sizes that are either local or metastatic, with stage IV indicating a BC of any size that has metastasized not only to local lymph nodes but to other body parts such as the bone (Cancer Australia, 2021; Chatterjee & McCaffrey, 2014; Giuliano, Edge, & Hortobagyi, 2018; National Breast Cancer *National Breast Cancer Foundation,* 2021). The staging of the cancer along with the cellular and tissue origin of the cancer and the defined molecular subtype culminate into a profile of diagnosis that informs the oncologist on how to treat the cancer and the best treatment regime for that patient.

2.2 Tumor microenvironment

The ECM plays a critical role in not only tissue homeostasis but also tumor development, as a diverse array of ECM molecules interact with other molecules but also directly with cells. Cytokines, GFs, hormone molecules, PGs and nutrients present within the ECM influence the development of BC and the expression of hallmarks. They can be used to support the pathological proliferation upregulation, provide a platform for angiogenesis and also assist in remodeling the microenvironment to better support the developing BC (Mannello, Ligi, & Canale, 2013; Nguyen-Ngoc et al., 2012).

In addition to the non-cellular characteristics of the tumor microenvironment, cells that are influenced by the primary cancer, epithelial in origin, alter not only their morphology and phenotype but their behavior to assist in facilitating the further propagation of the cancer. This can be seen upon normal mammary fibroblasts that undergo re-education by the cancer cells, turning them into cancer-associated fibroblasts (CAFs). CAFs are known to support tumor development and metastasis by inducing angiogenesis,

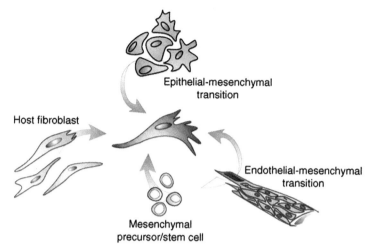

Fig. 3 Origin of CAFs. CAFs have a different site of origin, including local fibroblasts of the tissue and its mesenchymal precursors. They can also originate from epithelial and endothelial cells by undergoing an endothelial–mesenchymal transition. *Reproduced with permission from Östman, A., & Augsten, M. (2009). Cancer-associated fibroblasts and tumor growth–bystanders turning into key players.* Current Opinion in Genetics & Development, 19(1), 67–73.

ECM remodeling and recruitment of inflammatory cells. They produce and secrete molecules such as the small cytokine stromal-derived factor-1 (SDS-1) that drives angiogenesis by acting in a paracrine fashion *via* interactions with CXCR4. Similarly, hepatocyte growth factor (HGF) acts upon the c-MET receptor tyrosine kinase again in a paracrine fashion, which has shown to promote metastasis (Place, Huh, & Polyak, 2011; Tyan et al., 2011). Their sources of origin, as seen in Fig. 3, include local fibroblasts and their mesenchymal precursors, along with normal and malignant epithelial cells through epithelial–mesenchymal transition. In addition, the endothelial-mesenchymal transition has been shown to produce CAFs. The typical marker profile of CAFs includes platelet-derived growth factors receptor β (PDGFR- β); fibroblast activation protein (FAP); vitamin D receptor; and fibroblast-specific protein-1 (FSP1/S100A4). Additionally, CAFs express myofibroblast markers such as smooth muscle actin; laminin-1; and transforming growth factor-beta (TGF-β). As such, many CAF markers are not unique for this cell type (Herrera et al., 2013; Östman & Augsten, 2009; Santi, Kugeratski, & Zanivan, 2018). Clinically, the importance of CAFs are significant, as previous studies have shown that gene expression profiles of these cells are linked to poorer patient prognoses and outcomes (Eiro et al., 2018; Liao, Luo, Markowitz, Xiang, & Reisfeld, 2009; Santi et al., 2018).

Just as the BC can re-educate surrounding cells to support the progression and propagation of the cancer, they can also eradicate cells that may be competing for nutrients, room or behave as oncogenic suppression upon the tumor. Such cell populations include the myoepithelial cells which have been implicated in tumor suppressing behaviors by secreting protease inhibitors and downregulating matrix metalloprotease levels. This paracrine action by myoepithelial cells has a negative impact upon cell growth, invasion and angiogenic behaviors of BC cells. Myoepithelial cells differentiate from progenitor cell population that replace and maintain mammary homeostasis (Barsky & Karlin, 2005). However, it has been reported that there is a gradual loss of these progenitor cells over time within the tumor microenvironment. This, therefore, has a critical impact upon the replenishment of myoepithelial cell populations that decrease over time due to a lack of cell replenishment. Additionally, as the population decreases, subsequent tumor suppressing factors excreted by the myoepithelial population into the surrounding stroma decline, thereby providing a more favorable microenvironment for tumor progression. In fact, a criterion that differentiates an *in situ* carcinoma from an invasive carcinoma is the disappearance of the myoepithelial cell layer according to a paper published by Lerwill in the American Journal of Surgical Pathology in 2004 (Hu et al., 2008; Lerwill, 2004). The tumor microenvironment is ever-evolving and adapting, that displays retraining of host cells or a depletion of oncogenic suppressing cells that leads to a more hospitable tumor microenvironment (TME) and cancer progression.

Furthermore, the vascular network of the BC TME is an integral part of the development and progression, and mentioned previously, is partially induced by local and regional CAF populations. The blood and lymphatic vasculature networks supply oxygen and nutrition to the tumor site, similar to normal tissue along with carbon dioxide and metabolic waste removal. Both are crucial for tumor growth and sustenance and therefore, angiogenesis is an important feature of tumorigenesis. However, tumor blood vessels show aberrant structures and are often inefficient and leaky. This high permeability of the aberrant blood vessels is a result of an abnormal basement membrane and larger gaps between endothelial cells, which form the vessels. The dysregulation of angiogenesis in tumor tissue is based on an impaired expression profile of various important factors for blood vessel formation, such as vascular endothelial growth factor (VEGF) and basic fibroblast growth factor (FGF-2) (Hashizume et al., 2000; Less, Skalak, Sevick, & Jain, 1991; Munn, 2003).

The ECM plays an equally important role in BC, which is comprised of many non-cellular constituents. Proteoglycans have emerged as an important player in the regulation of BC but have also showed to be implicated in the disease's progression. PGs contain a core protein that can possess a variety of functional glycosaminoglycan (GAG) side chains. They are found in a number of locations and are typically classified by this, which include intra-cellular, cell-surface, pericellular and most importantly within the ECM (Clegg et al., 2020; Iozzo & Schaefer, 2015; Ruoslahti, 1988). Serglycin for instance promotes cell migration and invasion through anchorage-independent cellular movement, while PGs such as decorin and biglycan have been implicated in phenotypic conversion of cancer cells and cellular proliferation, respectively (Dawoody Nejad, Biglari, Annese, & Ribatti, 2017; Korpetinou et al., 2013; Recktenwald et al., 2012). Versican has also been implicated in BC, with a study by Ricciardelli et al., 2002, showing that lower levels of the ECM PG have been implicated in lower relapse rates, clinically implicating PGs as a potential therapeutic target and a prognostic indicator (Ricciardelli et al., 2002).

It is evident that BC is supported by an equally complex and heterogeneous ECM comprising of both cellular and non-cellular contributors that aid in its development and progression. There is a need to better understand the relationship between the primary tumor and surrounding TME in hopes of developing more effective treatments, prognostic and diagnostic tools and for potential drug screening applications.

2.3 Current treatments for BC

This heterogeneity of BC and the tendency to further mutate makes it difficult to effectively treat, especially taking into consideration the capacity of the disease to develop resistance to drug treatments, a phenomenon still not fully understood (Wind & Holen, 2011). A prognosis is a mixture of a predicted outcome of disease along with the chances of an individual to recover. A prognosis does not wholly depend upon the ability of the disease to mutate into a drug resistant cancer but also depends upon the localization, the stage of cancer and the subtype, along with the genome and epigenome of the individual, which the latter two factors are currently not routinely incorporated into prognostic measurements (Sun et al., 2017). As previously discussed, BC can upregulate specific markers such as estrogen or progesterone receptors, which in some incidences can be favorable towards a prognosis as they can be targeted by certain interventions. Oncologists often exploit these markers when prescribing antineoplastic drugs and have also been used in combination with other therapies to improve survival and reduce relapse. These are known as targeted or hormone therapies and target the markers upregulated by the BC (Masoud & Pagès, 2017).

Several risk factors can increase an individual for developing BC in their lifetime which includes lifestyle factors, family history, the presence of the BRCA1 and BRCA 2 genes and breast density. Women with high risks, such as having the BRCA1/2 genes and a high breast density, have the option to have preventative surgery known as prophylactic mastectomy which can greatly reduce their risk of developing BC. Women that have been diagnosed with local *in situ* BC can also have their breasts removed known as contralateral prophylactic mastectomy which will again lower their chances of reoccurrence. Furthermore, another form of preventative surgery that women can have a prophylactic oophorectomy which is the removal of their ovaries and usually their fallopian tubes. This greatly decreases the chance of ovarian cancer and some studies have indicated that it can also decrease their chances of developing BC, however, the verdict is still out on whether this is effective for at-risk individuals of developing BC (Giannakeas & Narod, 2018; National Cancer Institute, 2013).

Following a diagnosis, an oncologist may want to perform surgery, this is predominantly chosen for cases that have not developed metastatic BC. The treatment will also include drug treatment before or after the main treatment, that of the surgery. These are referred to as neoadjuvant or adjuvant therapies and will be performed depending upon tumor size, aggressiveness, invasive status and the age of the patient to name a few. Neoadjuvant chemotherapy will be performed to reduce the size of the tumor in an attempt to shrink the size, which may lead to more surgical options. Similarly, adjuvant treatment is predominantly chosen if the tumor is aggressive, invasive and has unfavorable prognostic factors. Ultimately, the oncologist will determine which treatment pathway will be best for a patient based on pathological assessments.

According to the American Cancer Society, two out of three BC will be hormone receptor-positive, meaning there is gene amplification for the ER and/or PR receptors that will increase the growth potential of the cancer. Hormone therapies, such as tamoxifen, can be prescribed to women that have ER+ BC which interfere with estrogen receptors in mammary epithelium, acting as antagonists to the receptor function, interestingly does the opposite to some other tissues such as bone and cardiovascular tissue, where it behaves as an estrogen. Because of this high selectivity of tamoxifen, it is further categorized as a selective estrogen receptor modulator (SERM) (American Cancer Society, 2020).

The HER2+ subtype displays an increase in proliferative and aggressive behaviors while having an increasing tendency to become invasive. New therapies that target the membrane receptor complex has led to an increase in positive prognostic outcomes for individuals diagnosed with HER2+ subtype. There are three drug therapy classes that target this subtype, known as monoclonal antibody-based, antibody-drug conjugates and kinase inhibitors. Trastuzumab and Pertuzumab, consumer names Herceptin and Perjeta respectively, are monoclonal antibodies that act similarly upon the HER2 receptor by binding to it, thereby antagonizing its function and inhibiting proliferation and survival capabilities of HER2-dependent tumors. These drugs have been used as neoadjuvant and adjuvant treatments as treatment in the absence of surgery shows lower success rates and increased relapses, which is why combination therapies when applicable are employed (Hudis, 2007). Antibody-drug conjugates such as ado-trastuzumab emtansine, also known as Kadcyla, is given as an adjuvant treatment for metastatic HER2 + BC. Conjugate therapies show great promise as they are highly selective and minimize systemic host tissue damage, while in the case of Kadcyla, the monoclonal antibody trastuzumab will target the HER2 complex on tumor cells, but also act as a delivery system for its conjugated cytotoxic drug, T-DMT1. The conjugated chemotherapeutic drug will then enter the cell through a process known as endocytosis and act upon the cytoskeleton inducing cell cycle arrest and activate apoptotic pathways (Barok, Joensuu, & Isola, 2014). Lastly, since the HER2 protein complex is part of the receptor tyrosine kinase family, drugs that antagonize and inhibit the ability of the HER2 kinase domain are referred to as kinase inhibitors (Wang et al., 2006).

BC not associated with an overexpression of estrogen, progesterone or HER2 protein complex are categorized as TNBC, as discussed previously. Since this subtype of BC does not display any of these molecular markers, no hormone or targeted therapies can be considered by the oncologist, and therefore the predominant treatment is chemotherapy and surgery. Patients diagnosed with stage 1–3 TNBC will likely be treated through surgery, along with neoadjuvant or adjuvant chemotherapies in an attempt to eradicate the tumor. Stage 4 TNBC patients will likely be given one or a mixture of platinum-based drugs such as cisplatin. In other cases, the oncologist will consider radiation therapy, Poly-(ADP-Ribose) Polymerase inhibitors (PARPs) or in some instances where the programmed cell death 1 ligand (PD-L1) is upregulated, then immunotherapy can be considered such as Atezolizumab (Beniey, Haque, & Hassan, 2019; Cerbelli et al., 2017; Wahba & El-Hadaad, 2015).

Despite emerging drugs coming to market such as retinoids, derivatives of vitamin A, and immunotherapies, therapeutics for cancer still have substantial shortcomings (Garattini et al., 2014; Williams et al., 2017). From significant side-effects to complete failure of drugs for patients, chemotherapeutics, radiotherapy and other treatments are predominantly blunt tools in the fight against BC. Molecular specific and hormone specific drugs, such as trastuzumab or tamoxifen aim to better treat BC patients without the widespread damage that can be caused by non-specific treatments (Ruddy et al., 2013). Precision medicine aims to treat patients at an individual level, incorporating other factors in treatment design such as age, sex and lifestyle. There has been a failure to translate this style of treatment into the clinic, and while there are some instances of precision screening employed such as the HercepTest that determines the presence of HER2 (Barrera-Saldaña, 2020), overall deployment of precision technologies have yet to be implemented.

More effort and resources within the scientific community and the healthcare system must be provided to move away from the generalized and blunt treatment of BC that produce both short-term and long-term health side effects and move into the development and deployment of precision modeling and medicines.

3. 3D tumor modeling
3.1 Three-dimensional (3D) models

For decades the study of BC and the development of novel treatments are typically performed in *in vitro* and *in vivo* models. Animal models, predominantly rodents such as mice and rats, have provided pre-clinically important information regarding the mammary microenvironment, both normal and pathological. Additionally, they are typically the steppingstone for novel drugs that then enter human clinical trials. To study cancer or to develop new oncology drugs, murine or even human cancer tumors and cells can be transplanted into immune-deficient mice (xenografts). Mice are more often used as models of BC over rats due to higher similarities with human mammary anatomy. Unfortunately, human and murine physiology is inherently different, in particular the immune system. In murine models of BC, often the murine mammary fat pad is used as a receptive microenvironment (Kuperwasser et al., 2004). Importantly, the mammary fat pad of the mouse does not resemble the anatomy of the human breast and thus requires prior humanization to enable studies on human epithelial cells (Wronski, Arendt, & Kuperwasser, 2015). In addition, murine models frequently tend to generate false-positive results, with drugs later failing in human trials. Adding to this, results and repetitions are also heterogeneous, primarily due to genetic differences, sex, applied methods, tumor sites, chosen time points and the biological activity of the tumors (DeRose et al., 2013; Vandamme, 2014).

Similarly, *in vitro* models such as the petri dish or other conventional two-dimensional (2D) culture models have also provided fundamental understanding of BC and has been involved at some stage in the development of BC treatments. These models possess favorable characteristics such as their high reproducibility, high throughput and simplicity. Disappointingly, they suffer from detrimental pitfalls that have contributed to a lack of progress in our understanding and contributed to novel treatment development that later falls prey to the widely known attrition rate of the pharmaceutical pipeline (Matthews, Hanison, & Nirmalan, 2016). The TME is a complex, multicellular three-dimensional (3D) environment that a monolayer single cell type population can never hope to re-encapsulate, which is the case of all 2D culture models (Imamura et al., 2015; Jensen & Teng, 2020).

These gaps presented by conventional use of 2D and animal models have provided researchers with opportunities to develop advanced 3D culture models that aim to demonstrate biomimetic mammary properties. 3D models have already been deployed for BC research and there are hopes for use within the clinical setting as drug screening, diagnostic and prognostic tools. These models can support more native cellular morphologies and 3D behaviors such as spheroid formation. Additionally, they aim to support complex cell-cell and cell-ECM interactions, leading to a more physiologically relevant understanding of BC that is summarized in Fig. 4 comparing 2D, animal and 3D models. Further, these advanced models may provide a platform for development of more sophisticated drugs that will also make it through clinical trials. Insights on precise cell-drug and drug-ECM interactions will contribute to the future success of novel drug development and is currently not possible when using 2D models (Breslin & O'Driscoll, 2016; Imamura et al., 2015; Lovitt, Shelper, & Avery, 2014).

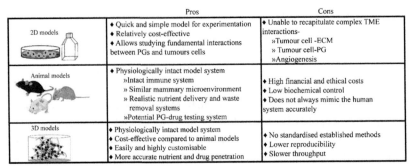

	Pros	Cons
2D models	♦ Quick and simple model for experimentation ♦ Relatively cost-effective ♦ Allows studying fundamental interactions between PGs and tumours cells	♦ Unable to recapitulate complex TME interactions- »Tumour cell -ECM » Tumour cell-PG »Angiogenesis
Animal models	♦ Physiologically intact model system »Intact immune system » Similar mammary microenvironment » Realistic nutrient delivery and waste removal systems »Potential PG-drug testing system	♦ High financial and ethical costs ♦ Low biochemical control ♦ Does not always mimic the human system accurately
3D models	♦ Physiologically intact model system ♦ Cost-effective compared to animal models ♦ Easily and highly customisable ♦ More accurate nutrient and drug penetration	♦ No standardised established methods ♦ Lower reproducibility ♦ Slower throughput

Fig. 4 2D and animal models both possess favorable characteristics and have been deployed for decades within the field of BC. Despite their many uses and fundamental contribution to our understanding of BC and treatment of the disease, they pertain significant pitfalls that have led researchers to develop 3D biomimetic models. *Reproduced from Clegg, J., Koch, M. K., Thompson, E. W., Haupt, L. M., Kalita-de Croft, P., & Bray, L. J. (2020). Three-dimensional models as a new frontier for studying the role of proteoglycans in the normal and malignant breast microenvironment.* Frontiers in Cell and Developmental Biology, 8, 1080.

Several 3D *in vitro* culture models have been developed to date. Such models include anchorage-independent methodologies that possess no substrate for cellular attachment. These models can include techniques such as coating surface or the use of low-attachment plates to support the formation of spheroids by cancer cells (Friedrich, Seidel, Ebner, & Kunz-Schughart, 2009; Ivascu & Kubbies, 2006). The spheroids that form from the deployment of these models present more physiological relevant features such as oxygen and nutrient gradients, heterogenous proliferation and cell–cell interactions. Additionally, matrix or scaffold based 3D models have also been developed (Fischbach et al., 2007; Kim, 2005). They support the growth of 3D structures and allow cell–ECM adherence, further, the use of cell suspension cultures that may use magnetic levitation or hanging drop method which supports cell–cell interactions, have been employed (Jaganathan et al., 2014; Tung, Hsiao, Allen, Ho, & Takayama, 2011).

Beyond cell–cell interactions, the influence of the ECM and therefore, cell–ECM interactions are equally important in both normal and pathophysiological tissue. This is one of the most significant disadvantages of 2D models, as they facilitate some cell–cell interactions, but they cannot hope to mimic any cell–ECM interactions. Several options are available for biomaterial-based 3D cell culture, with scaffold-based 3D culture methods being one example. Natural substrates used to mimic the ECM include basement membrane extracts from the Engelbreth–Holm–Swarm murine tumor (lrECM), also known as matrigel (Kenny et al., 2007; Thomas, Zeps, Rigby, & Hartmann, 2011), and collagens such as collagen I, which is often used in combination with fibrinogen (Carter, Gopsill, Gomm, Jones, & Grose, 2017; DeRose et al., 2013). Disadvantages associated with the use of natural matrices include high batch-to-batch variability, limited mechanical customizability and the process of substrate shrinkage. These factors can alter cellular signaling, leading to difficulties in the ability to control culture progression and degradation of the scaffold. Natural biomaterials are also limited in their possibility to mimic different/specific tissue environments (Kaemmerer, Garzon, Lock, Lovitt, & Avery, 2016; Kim, 2005; Langhans, 2018).

Fully synthetic matrices overcome these limitations, as they offer the greatest level of experimental control through defined, designed, and tunable material properties. Today, polyacrylamide (PA) and poly-ethylene glycol (PEG)-based synthetic matrices are used in cancer research. Scaffolds, made of polymer materials, such as electrospun fibers, provide pre-formed support for 3D culture and allow control of cell microenvironment. It is a stable construct suitable for long term experiments and shows a high degree of reproducibility. Due to its high stability, it is a suitable approach for use as implants or to mimic bone constructs for BC-bone metastasis research (Kaemmerer et al., 2016; Kim, 2005; Langhans, 2018). Examples include the use of highly micro-porous scaffolds made of poly(ε-caprolactone) (Rao et al., 2016), and poly-ether-urethane foam (Angeloni et al., 2017). However, visualization of integrated cells and recovery of viable cells can be hindered.

Hydrogels are considered one of the most successful 3D-ECM models developed thus far and are widely used. They provide cells with not only anchorage-sites for adherence but also a dynamic microenvironment consisting of ECM molecules such as PGs, GFs, cytokines and other soluble factors (Langhans, 2018). Hydrogels are classified into three types, natural, synthetic and semi-synthetic. Synthetic hydrogels are formulated using completely synthetic polymers and possess favorable characteristics such as high stiffness capabilities and a high degree of customizability. One such hydrogel are PA-based hydrogels that can mimic mechanoregulatory mechanisms that are a feature of the TME. Despite this, they do lack the ability of cell-induced remodeling due to their synthetic nature. Similarly, in poly(ethylene glycol) (PEG) hydrogels, the biochemical and mechanical properties are easily controlled and have shown potential to integrate cell-responsive degradation sites. Some synthetic hydrogels can be mineralized to mimic tissues such as bone. They include poly(glycolic) acid (PGA) and poly(lactic) acid, which are fabricated by gas foaming or particulate leaching and electrospinning. Mineralization of these hydrogels can be performed by incubating in simulated body fluid that forms an appetite layer or through incorporation of synthetically derived hydroxyapatite nanoparticles during the gel fabrication process (Infanger, Lynch, & Fischbach, 2013; Murphy, Kohn, & Mooney, 2000). Additional synthetic hydrogels employ polymers such as electrospun polycaprolactone (PCL), poly(acrylic acid) or polyvinyl alcohol–polyacrylic (Aswathy, Narendrakumar, & Manjubala, 2020).

Natural hydrogels are another subcategory of hydrogels that utilize naturally derived polymers as their architectural framework in contrast to synthetic polymers. PG-based hydrogels such as heparin or hyaluronic acid currently exist, along with the widely used matrigel and also gelatin fibrinogen and alginate-based hydrogels. A number of favorable features that are present include innate biocompatibility, inclusion of soluble factors or signaling molecules along with their ability to mimic softer tissues (Dhaliwal, 2012; Jensen & Teng, 2020). Unfortunately, they too possess characteristics such as uncontrollable degradation and are less customizable compared to synthetic hydrogels. Ultimately the type of gel that will be utilized will depend on the researcher, the tissue type that is being mimicked and the desired outcomes.

Semi-synthetic hydrogels aim to take the more advantageous characteristics of both natural and synthetic hydrogels and combine them into one hydrogel. Semi-synthetic hydrogels possess the high customizability seen in synthetic hydrogels, but also are biologically and biochemically more relevant as seen in natural hydrogels. This leads to more sophisticated cell cultures and allows for more complex experimentation. StarPEG–heparin and gelatin methylacrylol (GelMA) hydrogels as well as semi-synthetic superabsorbent polymer material-based gels such as cationic guar gum/poly (acrylic acid or chitosan cross-linked poly(acrylic acid) hydrogels are members of the semi-synthetic hydrogel classification and present properties of both natural and synthetic hydrogels (Bray et al., 2015; Mignon, De Belie, Dubruel, & Van Vlierberghe, 2019).

3.2 3D models for BC

Statistics from 2016 showed that the use of 3D culture techniques has increased over the last 10 years in the field of BC research (Rijal & Li, 2016), and has improved knowledge of cancer progression. By comparing results of 2D *versus* 3D cultures, many studies have shown genetic, phenotypic and proliferative variances of normal mammary gland epithelial cells (Roskelley, Srebrow, & Bissell, 1995; Zhong et al., 2014) and BC cells (Han et al., 2010; Kenny et al., 2007). Additionally, BC cells in 3D cultures show differences in chemotherapeutic sensitivity (Breslin & O'Driscoll, 2016; Lovitt, Shelper, & Avery, 2015). In BC research, naturally derived biomaterials for hydrogels, including matrigel, collagen, decellularized ECM and others (Lee & Mooney, 2001), are mostly used in comparison to synthetic biomaterials (Rijal & Li, 2016).

Several studies have used 3D culture methods to model the mammary gland or BC tumors *in vitro*. For example, Carter et al., 2017 isolated myoepithelial cells and luminal epithelial cells from isolated ductal organoids and combined the two primary cell types in a rat tail collagen I gel to recreate a bilayer glandular structure. Induction of HER2 expression initiated a luminal filling and thus imitated DCIS (Carter et al., 2017). Primary BC tissue has also been cultured in rat tail collagen gels to study its tamoxifen response (Leeper et al., 2012) and additionally, collagen I have been used as a platform for solid tumor development of cancer cell lines (Szot, Buchanan, Freeman, & Rylander, 2011). Thomas et al., 2011 used a pre-set bed of atrigel to culture epithelial cells, isolated from expressed human breastmilk, to generate human mammary alveolar structures (Thomas et al., 2011). The same culture method was used in studies correlating the morphologies of 3D cultured cancer cells to their gene expression profile (Han et al., 2010; Kenny et al., 2007). Furthermore, the resistance of primary BC cells and cell lines against targeted agents and antineoplastic drugs were tested using matrigel cell cultures (Li, Chow, & Mattingly, 2010; Lovitt et al., 2015). This commonly used 'on-top' culture method on matrigel is not entirely a 3D culture method. Although the cells grow on an ECM substrate and are able to infiltrate the matrix to a certain extent, they are not fully embedded and surrounded, and therefore, it is often referred to as a 2.5D culture (Langhans, 2018).

A study from 2015 by Mi and Xing showed that the phenotype of the BC cell line MDA-MB-435S, originally derived from metastatic ductal adenocarcinoma, could be reverted to a non-malignant state of mammary epithelial cells when cultured in a 3D self-assembling peptide nanofiber scaffold. This effect was not observed in cancer cells cultured in collagen or matrigel (Mi & Xing, 2015) demonstrating the influence of natural biomaterials on cancer cells. Furthermore, this study showed the advantage of synthetic biomaterials for 3D cultures in BC research to address specific mechanical questions of cell–ECM interactions. Previous work from our own research groups has demonstrated the use of PEG-heparin (Bray et al., 2015) and GelMA hydrogels (Taubenberger et al., 2016) for the culture of BC cell lines, such as MCF-7 and MDA-MB-231 (Loessner et al., 2016; Meinert, Theodoropoulos, Klein, Hutmacher, & Loessner, 2018). These studies, illustrated in Fig. 5, highlight the functionality of these semi-synthetic hydrogel models for the study of cell–matrix interactions *in vitro*.

The previously described advantages and disadvantages of natural biomaterials in comparison to synthetic and semi-synthetic approaches, in particular, the controllability of biochemical and biological characteristics and reproducibility should be taken into consideration for choosing a most suitable biomaterial for 3D cell culture modeling.

3.2.1 3D multicellular BC models

Apart from the importance of the biomaterial, the tissue microenvironment plays a crucial role in BC. Therefore, 3D culture techniques of mammary epithelial cells have now been extended to include other cell types or the intact tissue are used in culture.

Campbell, Davidenko, Caffarel, Cameron, & Watson, 2011 introduced a co-culture technique of murine mammary epithelial cells and adipocytes in a highly porous collagen-hyaluronic acid scaffold generated by a freeze-drying technique (Campbell et al., 2011). Holliday, Brouilette, Markert, Gordon, & Jones, 2009 published a model studying the interactions of BC cells and their microenvironment. In this study, pre-labeled BC cell lines and primary cells were cultured in rat tail collagen I together with normal or tumor-derived fibroblasts (Holliday et al., 2009). Another study generated heterogeneous breast tumors, consisting of BC cell lines and human pulmonary fibroblasts or patient-derived CAFs, by using the magnetic levitation system (Jaganathan et al., 2014). This model allowed cell–cell interactions between cancer and stromal cells, however, lacked the feature of cell–ECM interaction. Wang et al., 2010 co-cultured a mammary epithelial cell line simultaneously with a human mammary fibroblast cell line and human adipose-derived stem cells in a matrigel-collagen mix on a silk scaffold, combining two main cell types of mammary stromal tissue (Wang et al., 2010). This model provided a mammary stromal microenvironment which supports the formation of alveolar structures.

Other BC microenvironment studies have concentrated on the generation of a tumor angiogenesis model, in order to study the crucial role of

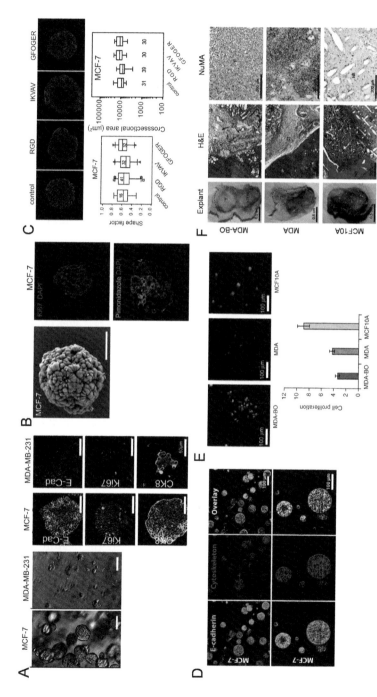

Fig. 5 See figure legend on next page.

blood vessel infiltration during tumor progression and metastasis and the evaluation of anti-angiogenesis therapies. Szot, Buchanan, Freeman, and Rylander (2013) published a bilayered, collagen I hydrogel, including BC cell lines in one layer and telomerase-immortalized human microvascular endothelial cells in a second layer with an acellular collagen layer in-between. This construct was maintained in a transwell plate, allowing cultivation in two specialized culture medium relevant to each cell type (Szot et al., 2013). Another BC angiogenesis study investigated cellular crosstalk using a hanging-drop co-culture technique, where the BC cell line was cultured in 2D and the microvascular endothelial cells were cultured in hanging culture inserts (Buchanan et al., 2012). In this type of angiogenesis model, the influence of both cell types is solely reduced to paracrine signaling and does not allow for direct cell-cell interactions. Furthermore, both cell types

---

Fig. 5 *In vitro* models of breast cancer in 3D hydrogels. (A) Light microscopy (scale bar: 100 μm) and immunofluorescent staining images (scale bar: 50 μm) of 3D tumor microenvironment of breast cancer cells MCF7 and MDA-MB-231 encapsulated in starPEG-heparin hydrogels (blue: Hoechst, red: phalloidin, green: target antibody). (B) SEM image of 3D MCF7 spheroids cultured in PEG-Heparin gels after 14 days of culture (scale bar: 50 μm). Immunofluorescent staining of the 3D MCF7 spheroids (with DAPI: *blue*) for Ki67 (*red*, top image) and pimonidazole (*green*, bottom image) to study proliferation and hypoxia respectively. (C) Effect of different ECM ligands (RGD, IKVAV, GFOGER) on MCF7 spheroid morphology and growth in PEG-Heparin gels was studied using immunofluorescent staining (phalloidin: *red*, DAPI: *blue*). Shape factor and cross-sectional area of the spheroids were quantified from confocal images. (D) Cellular morphology of MCF7 spheroids in GelMA-based hydrogels using immunofluorescent staining for E-cadherin (*green*), cytoskeleton (*red*), and nuclei (*blue*). (E) Analysis of viability of metastatic breast cancer cells (MDA-BO and MDA) and non-tumorigenic breast cells (MCF10a) in GelMA-based hydrogels using Live/Dead staining [Top image: FDA: live (*green*), PI: dead (*red*)] and AlamarBlue assay (Bottom image: mean ± SEM). (F) These cell-laden GelMA hydrogels were implanted in NOD-SCID mice with humanized tissue-engineered bone to study breast cancer invasion of the bone tissue. *Reproduced with permission from Panel (A) Bray, L. J., Binner, M., Holzheu, A., Friedrichs, J., Freudenberg, U., Hutmacher, D. W. et al. (2015). Multi-parametric hydrogels support 3D in vitro bioengineered microenvironment models of tumour angiogenesis. Biomaterials, 53, 609–620. Panels (B and C) Taubenberger, A. V., Bray, L. J., Haller, B., Shaposhnykov, A., Binner, M., Freudenberg, U. et al. (2016). 3D extracellular matrix interactions modulate tumour cell growth, invasion and angiogenesis in engineered tumour microenvironments. Acta Biomaterialia, 36, 73–85. Panel (D) Meinert, C., Theodoropoulos, C., Klein, T. J., Hutmacher, D. W., & Loessner, D. (2018). A method for prostate and breast cancer cell spheroid cultures using gelatin methacryloyl-based hydrogels. Prostate cancer (pp. 175–194). Springer. Panel (E) Loessner, D., Meinert, C., Kaemmerer, E., Martine, L. C., Yue, K., Levett, P. A. et al. (2016). Functionalization, preparation and use of cell-laden gelatin methacryloyl–based hydrogels as modular tissue culture platforms. Nature Protocols, 11(4), 727–746.*

are cultured in their optimal cell culture medium and are not exposed to the same cultivation conditions. Other studies have investigated the interaction of mammary epithelial cells and endothelial cells by co-culturing these cells on matrigel, allowing direct cell contact (Ingthorsson et al., 2010; Shekhar, Werdell, & Tait, 2000).

Furthermore, our research groups have previously used PEG-heparin hydrogels to cultivate co- and tri-cultures to investigate angiogenesis of BC (Fig. 6). The starPEG-heparin hydrogel, for example, utilizes the synthetic star-shaped PEG material crosslinked with the natural heparin, chemically crosslinked together by covalently bonding the carboxylic acid groups of heparin with the terminating amino groups of starPEG. We have demonstrated the capabilities within a metastatic breast to bone cancer context where we utilized BC cell lines in co-cultures or tri-cultures with human umbilical vein endothelial cells (HUVECs) and MSCs. This study not only further demonstrated the hydrogel's ability to culture multiple cell types at once but also provided understanding in the interactions between BC and primary human osteoblasts to better understand metastasis (Bray et al., 2018). Another study showed further tri-culture support for an angiogenic study that involved BC cell lines, LNCap and MCF-7, with HUVECs and MSCs that showed the development of tumor-vascular contacts and vascularization. This study also successfully demonstrated chemotherapeutic interactions that demonstrated less regression of the tumor and lower sensitivity of the cells in 3D compared to 2D (Bray et al., 2015).

3.2.2 3D primary cell and tissue cultures

Besides the cell type chosen for a model, its origin should be considered as well. While immortalized cell lines hold the advantage of being commercially available, it is known that they do not fully replicate the *in situ* tissue or primary cells (Al-Lamki, Bradley, & Pober, 2017; Burdall, Hanby, Lansdown, & Speirs, 2003). A study by Jaganathan et al. (2014) used the magnetic levitation method to create a scaffold-free culture model, which allows BC cells to interact with fibroblasts. Although they used several BC and fibroblasts cell lines to create heterogeneous tumors, they also used primary fibroblasts and CAFs and found that the heterogeneous tumors formed better with these cells than the fast-growing fibroblast cell lines (Jaganathan et al., 2014). Another example of using patient-derived fibroblasts with BC cell lines can be found in a fibroblasts-tumor cell interaction study published by Seidl, Huettinger, Knuechel, and Kunz-Schughart (2002). BC spheroids were co-cultured with fibroblast spheroids and thus

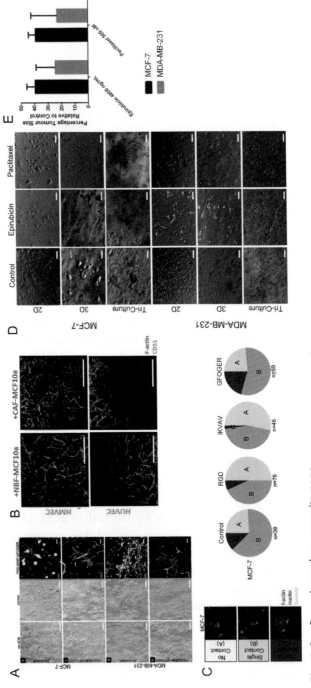

Fig. 6 See figure legend on opposite page.

were able to reflect tumor cell infiltration into the fibroblast population. They identified genes in the BC cells, which were differentially regulated upon contact with primary fibroblasts. Our most recent work used the PEG-heparin hydrogel model to explore endothelial-stromal interactions in a tissue-specific context (Koch et al., 2020). Patient-derived CAFs or normal fibroblasts from mammary tissue were co-cultured with either HUVECs or human mammary microvascular endothelial cells (Fig. 6B). Different effects of fibroblasts on the morphological characteristics of the vascular-like networks were visualized depending on the endothelial cell utilized.

In co-culture with primary cells in a 3D environment new insight into BC research can be gained. However, for more clinically relevant data or regenerative approaches, it is crucial to completely use primary cells. Showing a concept to regenerate autologous breast tissue, a research group

---

Fig. 6 3D multicellular models of breast cancer. (A) Brightfield images (scale bar: 200 µm) and immunofluorescence staining (*blue*: nuclei, *red*: actin, *green*: target anti-body, scale bar: 100 µm) of 3D tri-cultures of MCF7 and MDA-MB-231 cells with HUVECs and MSCs grown in PEG-based hydrogels that were co-cultured with hOBs in a 2D–3D co-culture model. (B) Analysis of 3D vessel network formation using 3D pro-jections of immunofluorescent-stained (*blue*: F-actin, *green*: CD31, scale bar: 500 µm) tri-cultures of MCF10a breast epithelial cells with normal breast fibroblasts (NBF)/cancer associated fibroblasts (CAF) and human mammary microvascular endothelial cells (HMVEC)/human umbilical vein endothelial cells (HUVEC). (C) Quantification of modu-latory effects of ECM ligands on interactions between endothelial networks and tumors using immunofluorescent staining (*blue*: nuclei, *red*: f-actin, *green*: CD31 or vWF) classified into three groups: no contact, single contact and internal contact. (D) Light microscopy images showing response of 2D, 3D (PEG-based hydrogel) and tri-culture (Tumor + HUVEC + MSC) models of breast cancer cells MCF7 and MDA-MB-231 cells to cytotoxic drug treatments (scale bar: 100 µm). (E) Regression of tumor size in response to drug treatment in 3D cultures of MCF7 and MDA-MB-231 in PEG-based hydrogels (mean ± STD). *Reproduced with permission from Panel (A) Bray, L. J., Secker, C., Murekatete, B., Sievers, J., Binner, M., Welzel, P. B. et al. (2018). Three-dimensional in vitro hydro-and cryogel-based cell-culture models for the study of breast-cancer metastasis to bone.* Cancers, 10(9), 292. *Panel (B) Koch, M. K., Jaeschke, A., Murekatete, B., Ravichandran, A., Tsurkan, M., Werner, C. et al. (2020). Stromal fibroblasts regulate microvascular-like network architecture in a bioengineered breast tumour angiogenesis model.* Acta Biomaterialia, 114, 256–269. *Panel (C) Taubenberger, A. V., Bray, L. J., Haller, B., Shaposhnykov, A., Binner, M., Freudenberg, U. et al. (2016). 3D extra-cellular matrix interactions modulate tumour cell growth, invasion and angiogenesis in engineered tumour microenvironments.* Acta Biomaterialia, 36, 73–85. *Panels (D and E) Bray, L. J., Binner, M., Holzheu, A., Friedrichs, J., Freudenberg, U., Hutmacher, D. W. et al. (2015). Multi-parametric hydrogels support 3D in vitro bioengineered microenvironment models of tumour angiogenesis.* Biomaterials, 53, 609–620.

isolated and co-cultured human mammary epithelial cells and preadipocytes from the same biopsy samples of normal breast tissue in a collagen matrix (Huss & Kratz, 2001) and showed that the cells grew in a tissue-like growth-pattern. In 2018, Wang et al. published a study of the effect of hypoxia on epithelial to mesenchymal transition (EMT) in a 3D hydrogel BC model (Wang et al., 2018). By using patient-derived isogenic primary and metastatic BC cell lines they showed that hypoxia decreased cell proliferation but enhanced EMT in these cells. Furthermore, by adding primary lung mesenchymal cells derived from the same patient, they were able to show the migration of these BC cells to the lung mesenchymal cells under hypoxic conditions (Wang et al., 2010). Another study presented a 3D *in vitro* model of the human breast duct bilayer, which reflects the human breast duct closely by isolating luminal and myoepithelial cells from reduction mammoplasties and combining them in a 3D matrix (Carter et al., 2017).

Tissue explant cultures are a good alternative to cell culture models for research of the cancer-microenvironment interaction or as drug testing platform. Explants have the advantage of having all cellular and non-cellular components of the original organ and thus display patient heterogeneity (Conde, Luvizotto, de Síbio, & Nogueira, 2012). For proper infiltration of the tissue with culture medium and oxygen, many studies use the slice culture method, where the breast tissue is cut into 200–300 µm thick slices and maintained in culture for a short period of up to 1 week. This method was used by Carranza-Torres et al. to screen for antineoplastic potential in natural compounds such as rosmarinic acid, caffeic acid, and ursolic acid (Carranza-Torres et al., 2015). Another group used 200 µm thick breast tumor slices in a short-term culture to study the anticancer effect of the drug taxol (van der Kuip et al., 2006). The slice culture method has the advantage of a controlled sample size and proper infiltration. However, they often disintegrate within a week of culture. Also, the proper infusion of tumor tissue with drugs is not mimicking the *in situ* situation of a tumor mass embedded in tissue. The complexity of a tumor also includes a gradient of oxygen and nutrients infiltrating the tumor forcing it to adapt (Stadler et al., 2015). This improper infiltration with drugs is one reason for drug resistance and should be included in drug screening approaches. To include the gradient of infiltration, thicker tissue explants are required as shown by Eigėlienė in 2006 (Eigeliene & Erkkola, 2006). In this study, the group investigated the effect of steroid hormones on the proliferative activity of mammary epithelial and stromal cells. The method is based on Trowell (1959) with the change that the explants were not fully immersed in the culture medium but placed on a metal grid. They showed that the explant cultures could be maintained for a culture period for up to 3 weeks and that apoptotic cells increase over the time of culture (Eigeliene & Erkkola, 2006).

4. Current limitations and considerations of 3D models for BC

To date, fundamental research has typically relied on conventional 2D cell culture methods and on small animal models to study the complex mechanisms of BC development. However, cell-cell and cell-matrix interactions during cancer progression in humans are not adequately reflected by these models. Furthermore, expensive *in vivo* models do not fully replicate human biology and can be difficult to monitor. Recently, there has been rapidly expanding research activities aimed at creating improved 3D *in vitro* models capable of mimicking vascular development *in situ* to study human tumor progression without animal harm or sacrifice (Chwalek, Tsurkan, Freudenberg, & Werner, 2014; Szot et al., 2013). There are three main requirements necessary for a more clinically relevant 3D model to mimic the tissue microenvironment. Firstly, the model requires an ECM substitute for mechanical support of the cellular structures. Secondly, different cell types need to be implemented, which allows direct and indirect communication between the different cell types. And lastly an adequate delivery of fluids, gases, and nutrients using a liquid medium (Kim, 2005).

Most 3D BC models rely on natural biomaterials, due to their high biocompatibility and cell growth support. However, these biomaterials also show high batch-to-batch variability, due to the inclusion of unknown components and their concentrations. In particular, the commonly used component matrigel is controversial, as it is derived from an Engelbreth-Holm-Swarm mouse sarcoma. Therefore, it resembles murine ECM, and additionally from a malignant microenvironment. Synthetic and semi-synthetic approaches overcome these limitations, however, often lack the complex tissue relevant microenvironment relevant to BC.

The complexity of models mimicking the *in situ* tissue can be modeled at various stages. The higher the complexity in modeling the original organ, the higher the biological relevance. However, the complexity of the model also decreases accessibility and can limit downstream analysis (Langhans, 2018). These facts should always be considered when choosing the origin of biological material for 3D cultures. In this study, different 3D cancer models for investigating cell–cell and cell–matrix interactions were developed, to mimic the mammary tissue microenvironment at different levels.

The vascular-like network formed by endothelial cells, as well as the epithelial mammospheres represent early stages in vessel and tumor formation. Advancements to the model could include the creation of a microenvironment with further matured cellular structures, for example, perfusable vascular networks and advanced tumor formations. To use further matured tumors, the cell culture needs to be either longer than 1 week or they could be precultured and included as already formed mammospheres of a certain size. The maturation of the vascular-like network may require supporting cells such as mesenchymal stem cells or pericytes, supporting the endothelial cells. Another approach would be to create a stronger diffusion of the medium with vital nutrients, by generating a gradient or mechanical movement. Also, the optimization of this culture for microfluidic devices could help in generating matured vessel-like structures, although accessibility for downstream analysis would be further reduced. Besides the further development of cellular components, other important cell types of the mammary or tumor microenvironment could be added such as immune cells or adipocytes because they, too, play an important role in BC development. Furthermore, the patient specificity could be increased by using patient-derived epithelial or BC cells.

Other considerations, apart from the appropriate biomaterial, should include the type and source of cells used in modeling the TME. At present, most studies are performed with cells readily obtainable and maintainable in culture, including immortalized cancer cell lines, fibroblasts as representative stromal cells, and HUVEC for generating a vascular-like structure. However, studies using patient-derived tissue, and therefore clinically relevant cell types, are still lacking. To fully develop insights into the mechanisms driving BC progression, an adequate and feasible model is critical for biological relevance, cost-effectiveness, and user-friendly sample handling.

5. Future prospective of 3D BC modeling systems

The research and clinical issues surrounding BC have limited our understanding and the treatment of the disease. Current treatment options and screening tools within the clinical settings have so far improved patient survival rates over the last few decades, but thousands of women worldwide still succumb to the disease. The way we treat and diagnose BC is a reflection of our understanding, and even though 2D and animal models have brought us here thus far, there are many underlying mechanisms to be fully elucidated.

The 3D studies discussed show the state-of-the-art biomaterials derived from several materials and methodologies and outline potential tissue engineering strategies for reconstructing the mammary microenvironment. All these 3D models aim to provide a biomimetic platform for deeper understanding of this disease such as investigating the role of angiogenesis in metastasis and its initial development within the TME as described by researchers such as Bray et al. (2015) and Szot et al. (2013). Additionally, these 3D models have already been shown to support complex metastatic experimentation as outlined by Bray et al. (2018) and Angeloni et al. (2017). Exploiting the customizability of many of the 3D models will allow researchers to manipulate properties like biochemical, physicochemical and mechanical properties that are more suitable for different stiffnesses and signaling processes present during different periods of BC development and during the remodeling processes of the TME. This will lead to a deeper understanding of the underlying mechanisms that contribute to the hallmarks displayed by BC and hopefully translate to more sophisticated treatments.

Incorporating patient tissue and primary BC cells are a fundamental requirement of translating many of these models into clinical tools or to validate them as mammary biomimetic tools. We currently employ xenograph techniques into animal models which have the ability to sustain these patient cancer tissues, but being able to support them *in vitro* will facilitate a more robust research paradigm that will influence future understanding and become critical players in the development of sorely needed novel drugs and treatment methods. As previously discussed, treatments regimes are currently developed for patients by a multi-disciplinary team of clinicians based upon the staging and molecular subtype determined. For HR-positive BC, we are able to provide targeted therapies such as tamoxifen that antagonized estrogen receptors and retards tumor growth. HER2-positive cancers can also be given targeted therapies such as trastuzumab and Kadcyla. Despite these innovations, TNBC remains to be the biggest challenge for the development of targeted treatments as there are currently known targets and therefore no targeted treatments on the market. Additionally, radiotherapy that employs ionizing radiation to tumor sites has been effective yet a cocktail of side-effects do present in patients due to off-target cellular damage and local and systemic reactive oxygen species injuring non-pathological tissues.

Just like side-effects presented by radiotherapy, chemotherapy and even targeted therapies present undesirable side-effects and thus remain a challenge for clinicians.

Future 3D models will be able to assist in the development of more sophisticated drugs that will present a reduced side-effect profile and will be more targeted and effective. Furthermore, it is hoped that future research will lead to targeted therapies for the TNBC subtypes of BC which will significantly increase survival rates of patients diagnosed with this subtype of BC. We anticipate that the future of research and clinical fields of BC will involve 3D culture models described here today that will play a myriad of roles in our pursuit of understanding and treatment of this complex cancer.

References

Al-Lamki, R. S., Bradley, J. R., & Pober, J. S. (2017). Human organ culture: Updating the approach to bridge the gap from in vitro to in vivo in inflammation, cancer, and stem cell biology. *Frontiers in Medicine, 4,* 148.

Allred, D. C. (2010). Ductal carcinoma in situ: Terminology, classification, and natural history. *Journal of the National Cancer Institute Monographs, 2010*(41), 134–138.

American Cancer Society. (2020). *Hormone therapy for breast cancer.* https://www.cancer.org/cancer/breast-cancer/treatment/hormone-therapy-for-breast-cancer.html.

Angeloni, V., Contessi, N., De Marco, C., Bertoldi, S., Tanzi, M. C., Daidone, M. G., et al. (2017). Polyurethane foam scaffold as in vitro model for breast cancer bone metastasis. *Acta Biomaterialia, 63,* 306–316.

Aswathy, S., Narendrakumar, U., & Manjubala, I. (2020). Commercial hydrogels for biomedical applications. *Heliyon, 6*(4), e03719.

Barok, M., Joensuu, H., & Isola, J. (2014). Trastuzumab emtansine: Mechanisms of action and drug resistance. *Breast Cancer Research, 16*(2), 1–12.

Barrera-Saldaña, H. A. (2020). Origin of personalized medicine in pioneering, passionate, genomic research. *Genomics, 112*(1), 721–728.

Barresi, R., & Campbell, K. P. (2006). Dystroglycan: From biosynthesis to pathogenesis of human disease. *Journal of Cell Science, 119*(2), 199–207.

Barsky, S. H., & Karlin, N. J. (2005). Myoepithelial cells: Autocrine and paracrine suppressors of breast cancer progression. *Journal of Mammary Gland Biology and Neoplasia, 10*(3), 249–260.

Beniey, M., Haque, T., & Hassan, S. (2019). Translating the role of PARP inhibitors in triple-negative breast cancer. *Oncoscience, 6*(1–2), 287.

Blows, F. M., Driver, K. E., Schmidt, M. K., Broeks, A., Van Leeuwen, F. E., Wesseling, J., et al. (2010). Subtyping of breast cancer by immunohistochemistry to investigate a relationship between subtype and short and long term survival: A collaborative analysis of data for 10,159 cases from 12 studies. *PLoS Medicine, 7*(5), e1000279.

Bray, L. J., Binner, M., Holzheu, A., Friedrichs, J., Freudenberg, U., Hutmacher, D. W., et al. (2015). Multi-parametric hydrogels support 3D in vitro bioengineered microenvironment models of tumour angiogenesis. *Biomaterials, 53,* 609–620.

Bray, L. J., Secker, C., Murekatete, B., Sievers, J., Binner, M., Welzel, P. B., et al. (2018). Three-dimensional in vitro hydro-and cryogel-based cell-culture models for the study of breast-cancer metastasis to bone. *Cancers, 10*(9), 292.

Breslin, S., & O'Driscoll, L. (2016). The relevance of using 3D cell cultures, in addition to 2D monolayer cultures, when evaluating breast cancer drug sensitivity and resistance. *Oncotarget, 7*(29), 45745.

Buchanan, C. F., Szot, C. S., Wilson, T. D., Akman, S., Metheny-Barlow, L. J., Robertson, J. L., et al. (2012). Cross-talk between endothelial and breast cancer cells regulates reciprocal expression of angiogenic factors in vitro. *Journal of Cellular Biochemistry, 113*(4), 1142–1151.

Burdall, S. E., Hanby, A. M., Lansdown, M. R., & Speirs, V. (2003). Breast cancer cell lines: Friend or foe? *Breast Cancer Research, 5*(2), 1–7.

Campbell, J. J., Davidenko, N., Caffarel, M. M., Cameron, R. E., & Watson, C. J. (2011). A multifunctional 3D co-culture system for studies of mammary tissue morphogenesis and stem cell biology. *PLoS One, 6*(9), e25661.

Cancer Australia. (2021). *Stages of breast cancer*. Australian Government. https://www.canceraustralia.gov.au/affected-cancer/cancer-types/breast-cancer/diagnosis/stages-breast-cancer.

Cancer Council. (2019). *Understanding breast cancer*. Cancer Council. https://www.cancer.org.au/assets/pdf/understanding-breast-cancer-booklet.

Carey, L. A., Perou, C. M., Livasy, C. A., Dressler, L. G., Cowan, D., Conway, K., et al. (2006). Race, breast cancer subtypes, and survival in the Carolina Breast Cancer Study. *JAMA, 295*(21), 2492–2502.

Carranza-Torres, I. E., Guzmán-Delgado, N. E., Coronado-Martínez, C., Bañuelos-García, J. I., Viveros-Valdez, E., Morán-Martínez, J., et al. (2015). Organotypic culture of breast tumor explants as a multicellular system for the screening of natural compounds with antineoplastic potential. *BioMed Research International, 2015*, 618021.

Carter, E. P., Gopsill, J. A., Gomm, J. J., Jones, J. L., & Grose, R. P. (2017). A 3D in vitro model of the human breast duct: A method to unravel myoepithelial-luminal interactions in the progression of breast cancer. *Breast Cancer Research, 19*(1), 1–10.

Cerbelli, B., Pernazza, A., Botticelli, A., Fortunato, L., Monti, M., Sciattella, P., et al. (2017). PD-L1 expression in TNBC: A predictive biomarker of response to neoadjuvant chemotherapy? *BioMed Research International, 2017*, 1750925.

Chatterjee, S. J., & McCaffrey, L. (2014). Emerging role of cell polarity proteins in breast cancer progression and metastasis. *Breast Cancer: Targets and Therapy, 6*, 15.

Chwalek, K., Tsurkan, M. V., Freudenberg, U., & Werner, C. (2014). Glycosaminoglycan-based hydrogels to modulate heterocellular communication in in vitro angiogenesis models. *Scientific Reports, 4*(4414), 8. https://doi.org/10.1038/srep0441.

Clegg, J., Koch, M. K., Thompson, E. W., Haupt, L. M., Kalita-de Croft, P., & Bray, L. J. (2020). Three-dimensional models as a new frontier for studying the role of proteoglycans in the normal and malignant breast microenvironment. *Frontiers in Cell and Development Biology, 8*, 1080.

Conde, S. J., Luvizotto, R. D. A. M., de Síbio, M. T., & Nogueira, C. R. (2012). Human breast tumor slices as an alternative approach to cell lines to individualize research for each patient. *European Journal of Cancer Prevention, 21*(4), 333–335.

Dawoody Nejad, L., Biglari, A., Annese, T., & Ribatti, D. (2017). Recombinant fibromodulin and decorin effects on NF-κB and TGFβ1 in the 4T1 breast cancer cell line. *Oncology Letters, 13*(6), 4475–4480.

DeRose, Y. S., Gligorich, K. M., Wang, G., Georgelas, A., Bowman, P., Courdy, S. J., et al. (2013). Patient-derived models of human breast cancer: Protocols for in vitro and in vivo applications in tumor biology and translational medicine. *Current Protocols in Pharmacology, 60*(1), 14.23.11–14.23.43.

Dhaliwal, A. (2012). 3D Cell culture: A review. *Material Methods, 2*, 162.

Eigeliene, N. H. P., & Erkkola, R. (2006). Effects of estradiol and medroxyprogesterone acetate on morphology, proliferation and apoptosis of human breast tissue in organ cultures. *BMC Cancer, 6*, 246.

Eiro, N., Gonzalez, L., Martinez-Ordonez, A., Fernandez-Garcia, B., Gonzalez, L. O., Cid, S., et al. (2018). Cancer-associated fibroblasts affect breast cancer cell gene expression, invasion and angiogenesis. *Cellular Oncology*. https://doi.org/10.1007/s13402-018-0371-y.

Fischbach, C., Chen, R., Matsumoto, T., Schmelzle, T., Brugge, J. S., Polverini, P. J., et al. (2007). Engineering tumors with 3D scaffolds. *Nature Methods*, *4*(10), 855–860.

Friedrich, J., Seidel, C., Ebner, R., & Kunz-Schughart, L. A. (2009). Spheroid-based drug screen: Considerations and practical approach. *Nature Protocols*, *4*(3), 309–324.

Garattini, E., Bolis, M., Garattini, S. K., Fratelli, M., Centritto, F., Paroni, G., et al. (2014). Retinoids and breast cancer: From basic studies to the clinic and back again. *Cancer Treatment Reviews*, *40*(6), 739–749.

Geddes, D. T. (2007). Inside the lactating breast: The latest anatomy research. *Journal of Midwifery & Women's Health*, *52*(6), 556–563.

Giannakeas, V., & Narod, S. A. (2018). The expected benefit of preventive mastectomy on breast cancer incidence and mortality in BRCA mutation carriers, by age at mastectomy. *Breast Cancer Research and Treatment*, *167*(1), 263–267.

Giuliano, A. E., Edge, S. B., & Hortobagyi, G. N. (2018). Of the AJCC cancer staging manual: Breast cancer. *Annals of Surgical Oncology*, *25*(7), 1783–1785.

Han, J., Chang, H., Giricz, O., Lee, G. Y., Baehner, F. L., Gray, J. W., et al. (2010). Molecular predictors of 3D morphogenesis by breast cancer cell lines in 3D culture. *PLoS Computational Biology*, *6*(2), e1000684.

Hashizume, H., Baluk, P., Morikawa, S., McLean, J. W., Thurston, G., Roberge, S., et al. (2000). Openings between defective endothelial cells explain tumor vessel leakiness. *The American Journal of Pathology*, *156*(4), 1363–1380.

Hassiotou, F., & Geddes, D. (2013). Anatomy of the human mammary gland: Current status of knowledge. *Clinical Anatomy*, *26*(1), 29–48.

Herrera, M., Islam, A. B., Herrera, A., Martín, P., García, V., Silva, J., et al. (2013). Functional heterogeneity of cancer-associated fibroblasts from human colon tumors shows specific prognostic gene expression signature. *Clinical Cancer Research*, *19*(21), 5914–5926.

Holliday, D. L., Brouilette, K. T., Markert, A., Gordon, L. A., & Jones, J. L. (2009). Novel multicellular organotypic models of normal and malignant breast: Tools for dissecting the role of the microenvironment in breast cancer progression. *Breast Cancer Research*, *11*(1), 1–11.

Hu, M., Yao, J., Carroll, D. K., Weremowicz, S., Chen, H., Carrasco, D., et al. (2008). Regulation of in situ to invasive breast carcinoma transition. *Cancer Cell*, *13*(5), 394–406.

Hudis, C. A. (2007). Trastuzumab—Mechanism of action and use in clinical practice. *New England Journal of Medicine*, *357*(1), 39–51.

Huss, F. R., & Kratz, G. (2001). Mammary epithelial cell and adipocyte co-culture in a 3-D matrix: The first step towards tissue-engineered human breast tissue. *Cells, Tissues, Organs*, *169*(4), 361–367.

Imamura, Y., Mukohara, T., Shimono, Y., Funakoshi, Y., Chayahara, N., Toyoda, M., et al. (2015). Comparison of 2D-and 3D-culture models as drug-testing platforms in breast cancer. *Oncology Reports*, *33*(4), 1837–1843.

Infanger, D. W., Lynch, M. E., & Fischbach, C. (2013). Engineered culture models for studies of tumor-microenvironment interactions. *Annual Review of Biomedical Engineering*, *15*, 29–53.

Ingthorsson, S., Sigurdsson, V., Fridriksdottir, A. J., Jonasson, J. G., Kjartansson, J., Magnusson, M. K., et al. (2010). Endothelial cells stimulate growth of normal and cancerous breast epithelial cells in 3D culture. *BMC Research Notes*, *3*(1), 1–12.

Iozzo, R. V., & Schaefer, L. (2015). Proteoglycan form and function: A comprehensive nomenclature of proteoglycans. *Matrix Biology*, *42*, 11–55.

Ivascu, A., & Kubbies, M. (2006). Rapid generation of single-tumor spheroids for high-throughput cell function and toxicity analysis. *Journal of Biomolecular Screening, 11*(8), 922–932.

Jaganathan, H., Gage, J., Leonard, F., Srinivasan, S., Souza, G. R., Dave, B., et al. (2014). Three-dimensional in vitro co-culture model of breast tumor using magnetic levitation. *Scientific Reports, 4*(1), 1–9.

Jensen, C., & Teng, Y. (2020). Is it time to start transitioning from 2D to 3D cell culture? *Frontiers in Molecular Biosciences, 7*, 33.

Jolicoeur, F. (2005). Intrauterine breast development and the mammary myoepithelial lineage. *Journal of Mammary Gland Biology and Neoplasia, 10*(3), 199–210.

Jolicoeur, F., Gaboury, L. A., & Oligny, L. L. (2003). Basal cells of second trimester fetal breasts: Immunohistochemical study of myoepithelial precursors. *Pediatric and Developmental Pathology, 6*(5), 398–413.

Joshi, P. A., Di Grappa, M. A., & Khokha, R. (2012). Active allies: Hormones, stem cells and the niche in adult mammopoiesis. *Trends in Endocrinology and Metabolism, 23*(6), 299–309.

Kaemmerer, E., Garzon, T. E. R., Lock, A. M., Lovitt, C. J., & Avery, V. M. (2016). Innovative in vitro models for breast cancer drug discovery. *Drug Discovery Today: Disease Models, 21*, 11–16.

Kenny, P. A., Lee, G. Y., Myers, C. A., Neve, R. M., Semeiks, J. R., Spellman, P. T., et al. (2007). The morphologies of breast cancer cell lines in three-dimensional assays correlate with their profiles of gene expression. *Molecular Oncology, 1*(1), 84–96.

Kim, J. B. (2005). Three-dimensional tissue culture models in cancer biology. *Seminars in Cancer Biology, 15*(5), 365–377. https://doi.org/10.1016/j.semcancer.2005.05.002.

Koch, M. K., Jaeschke, A., Murekatete, B., Ravichandran, A., Tsurkan, M., Werner, C., et al. (2020). Stromal fibroblasts regulate microvascular-like network architecture in a bioengineered breast tumour angiogenesis model. *Acta Biomaterialia, 114*, 256–269.

Korpetinou, A., Skandalis, S. S., Moustakas, A., Happonen, K. E., Tveit, H., Prydz, K., et al. (2013). Serglycin is implicated in the promotion of aggressive phenotype of breast cancer cells. *PLoS One, 8*(10), e78157.

Kuperwasser, C., Chavarria, T., Wu, M., Magrane, G., Gray, J. W., Carey, L., et al. (2004). Reconstruction of functionally normal and malignant human breast tissues in mice. *Proceedings of the National Academy of Sciences, 101*(14), 4966–4971.

Langhans, S. A. (2018). Three-dimensional in vitro cell culture models in drug discovery and drug repositioning. *Frontiers in Pharmacology, 9*, 6.

Lee, K. Y., & Mooney, D. J. (2001). Hydrogels for tissue engineering. *Chemical Reviews, 101*(7), 1869–1880.

Leeper, A. D., Farrell, J., Williams, L. J., Thomas, J. S., Dixon, J. M., Wedden, S. E., et al. (2012). Determining tamoxifen sensitivity using primary breast cancer tissue in collagen-based three-dimensional culture. *Biomaterials, 33*(3), 907–915.

Lerwill, M. F. (2004). Current practical applications of diagnostic immunohistochemistry in breast pathology. *The American Journal of Surgical Pathology, 28*(8), 1076–1091.

Less, J. R., Skalak, T. C., Sevick, E. M., & Jain, R. K. (1991). Microvascular architecture in a mammary carcinoma: Branching patterns and vessel dimensions. *Cancer Research, 51*(1), 265–273.

Li, Q., Chow, A. B., & Mattingly, R. R. (2010). Three-dimensional overlay culture models of human breast cancer reveal a critical sensitivity to mitogen-activated protein kinase kinase inhibitors. *Journal of Pharmacology and Experimental Therapeutics, 332*(3), 821–828.

Liao, D., Luo, Y., Markowitz, D., Xiang, R., & Reisfeld, R. A. (2009). Cancer associated fibroblasts promote tumor growth and metastasis by modulating the tumor immune microenvironment in a 4T1 murine breast cancer model. *PLoS One, 4*(11), e7965.

Loessner, D., Meinert, C., Kaemmerer, E., Martine, L. C., Yue, K., Levett, P. A., et al. (2016). Functionalization, preparation and use of cell-laden gelatin methacryloyl–based hydrogels as modular tissue culture platforms. *Nature Protocols, 11*(4), 727–746.

Lovitt, C. J., Shelper, T. B., & Avery, V. M. (2014). Advanced cell culture techniques for cancer drug discovery. *Biology*, *3*(2), 345–367.

Lovitt, C. J., Shelper, T. B., & Avery, V. M. (2015). Evaluation of chemotherapeutics in a three-dimensional breast cancer model. *Journal of Cancer Research and Clinical Oncology*, *141*(5), 951–959.

Macias, H., & Hinck, L. (2012). Mammary gland development. *Wiley Interdisciplinary Reviews: Developmental Biology*, *1*(4), 533–557.

Mannello, F., Ligi, D., & Canale, M. (2013). Aluminium, carbonyls and cytokines in human nipple aspirate fluids: Possible relationship between inflammation, oxidative stress and breast cancer microenvironment. *Journal of Inorganic Biochemistry*, *128*, 250–256.

Masoud, V., & Pagès, G. (2017). Targeted therapies in breast cancer: New challenges to fight against resistance. *World Journal of Clinical Oncology*, *8*(2), 120.

Matthews, H., Hanison, J., & Nirmalan, N. (2016). "Omics"-informed drug and biomarker discovery: Opportunities, challenges and future perspectives. *Proteome*, *4*(3), 28.

Medina, D. (1996). The mammary gland: A unique organ for the study of development and tumorigenesis. *Journal of Mammary Gland Biology and Neoplasia*, *1*(1), 5–19.

Meinert, C., Theodoropoulos, C., Klein, T. J., Hutmacher, D. W., & Loessner, D. (2018). A method for prostate and breast cancer cell spheroid cultures using gelatin methacryloyl-based hydrogels. In *Prostate Cancer* (pp. 175–194). Springer.

Mi, K., & Xing, Z. (2015). CD44 + /CD24 − breast cancer cells exhibit phenotypic reversion in three-dimensional self-assembling peptide RADA16 nanofiber scaffold. *International Journal of Nanomedicine*, *10*, 3043.

Mignon, A., De Belie, N., Dubruel, P., & Van Vlierberghe, S. (2019). Superabsorbent polymers: A review on the characteristics and applications of synthetic, polysaccharide-based, semi-synthetic and 'smart'derivatives. *European Polymer Journal*, *117*, 165–178.

Morgan, M. R., Humphries, M. J., & Bass, M. D. (2007). Synergistic control of cell adhesion by integrins and syndecans. *Nature Reviews Molecular Cell Biology*, *8*(12), 957–969.

Munn, L. L. (2003). Aberrant vascular architecture in tumors and its importance in drug-based therapies. *Drug Discovery Today*, *8*(9), 396–403.

Murphy, W. L., Kohn, D. H., & Mooney, D. J. (2000). Growth of continuous bonelike mineral within porous poly (lactide-co-glycolide) scaffolds in vitro. *Journal of Biomedical Materials Research: An Official Journal of the Society for Biomaterials and the Japanese Society for Biomaterials*, *50*(1), 50–58.

Muschler, J., & Streuli, C. H. (2010). Cell–matrix interactions in mammary gland development and breast cancer. *Cold Spring Harbor Perspectives in Biology*, *2*(10), a003202.

Staging & types of breast cancer. (2021). National Breast Cancer Foundation. https://nbcf.org.au/about-breast-cancer/diagnosis/stages-of-breast-cancer/.

National Cancer Institute. (2013). *Surgery to reduce the risk of breast cancer*. https://www.cancer.gov/types/breast/risk-reducing-surgery-fact-sheet.

Nguyen-Ngoc, K.-V., Cheung, K. J., Brenot, A., Shamir, E. R., Gray, R. S., Hines, W. C., et al. (2012). ECM microenvironment regulates collective migration and local dissemination in normal and malignant mammary epithelium. *Proceedings of the National Academy of Sciences*, *109*(39), E2595–E2604.

Östman, A., & Augsten, M. (2009). Cancer-associated fibroblasts and tumor growth—bystanders turning into key players. *Current Opinion in Genetics & Development*, *19*(1), 67–73.

Patel, V. N., Knox, S. M., Likar, K. M., Lathrop, C. A., Hossain, R., Eftekhari, S., et al. (2007). Heparanase cleavage of perlecan heparan sulfate modulates FGF10 activity during ex vivo submandibular gland branching morphogenesis. *Development*, *134*(23), 4177–4186.

PDQ Adult Treatment Editorial Board. (2020). Breast Cancer treatment during pregnancy (PDQ®). In *PDQ Cancer Information Summaries [Internet]* National Cancer Institute (US).

Perou, C. M., Sørlie, T., Eisen, M. B., Van De Rijn, M., Jeffrey, S. S., Rees, C. A., et al. (2000). Molecular portraits of human breast tumours. *Nature, 406*(6797), 747–752.

Place, A. E., Huh, S. J., & Polyak, K. (2011). The microenvironment in breast cancer progression: Biology and implications for treatment. *Breast Cancer Research, 13*(6), 1–11.

Ramsay, D., Kent, J., Hartmann, R., & Hartmann, P. (2005). Anatomy of the lactating human breast redefined with ultrasound imaging. *Journal of Anatomy, 206*(6), 525–534.

Rao, S. S., Bushnell, G. G., Azarin, S. M., Spicer, G., Aguado, B. A., Stoehr, J. R., et al. (2016). Enhanced survival with implantable scaffolds that capture metastatic breast cancer cells in vivo. *Cancer Research, 76*(18), 5209–5218.

Recktenwald, C. V., Leisz, S., Steven, A., Mimura, K., Müller, A., Wulfänger, J., et al. (2012). HER-2/neu-mediated down-regulation of biglycan associated with altered growth properties. *Journal of Biological Chemistry, 287*(29), 24320–24329.

Ricciardelli, C., Brooks, J. H., Suwiwat, S., Sakko, A. J., Mayne, K., Raymond, W. A., et al. (2002). Regulation of stromal versican expression by breast cancer cells and importance to relapse-free survival in patients with node-negative primary breast cancer. *Clinical Cancer Research, 8*(4), 1054–1060.

Rijal, G., & Li, W. (2016). 3D scaffolds in breast cancer research. *Biomaterials, 81*, 135–156.

Roskelley, C. D., Srebrow, A., & Bissell, M. J. (1995). A hierarchy of ECM-mediated signalling regulates tissue-specific gene expression. *Current Opinion in Cell Biology, 7*(5), 736–747.

Ruddy, K. J., Desantis, S. D., Gelman, R. S., Wu, A. H., Punglia, R. S., Mayer, E. L., et al. (2013). Personalized medicine in breast cancer: Tamoxifen, endoxifen, and CYP2D6 in clinical practice. *Breast Cancer Research and Treatment, 141*(3), 421–427.

Ruoslahti, E. (1988). Structure and biology of proteoglycans. *Annual Review of Cell Biology, 4*(1), 229–255.

Santi, A., Kugeratski, F. G., & Zanivan, S. (2018). Cancer associated fibroblasts: The architects of stroma remodeling. *Proteomics, 18*(5–6), 1700167.

Seidl, P., Huettinger, R., Knuechel, R., & Kunz-Schughart, L. A. (2002). Three-dimensional fibroblast–tumor cell interaction causes downregulation of RACK1 mRNA expression in breast cancer cells in vitro. *International Journal of Cancer, 102*(2), 129–136.

Shekhar, M. P., Werdell, J., & Tait, L. (2000). Interaction with endothelial cells is a prerequisite for branching ductal-alveolar morphogenesis and hyperplasia of preneoplastic human breast epithelial cells: Regulation by estrogen. *Cancer Research, 60*(2), 439–449.

Sinn, H.-P., & Kreipe, H. (2013). A brief overview of the WHO classification of breast tumors. *Breast Care, 8*(2), 149–154.

Stadler, M. W. S., Walzl, A., Kramer, N., Unger, C., Scherzer, M., Unterleuthner, D., et al. (2015). Increased complexity in carcinomas: Analyzing and modeling the interaction of human cancer cells with their microenvironment. In *Vol. 35. Seminars in cancer biology* (pp. 107–124). Academic Press.

Sun, Y.-S., Zhao, Z., Yang, Z.-N., Xu, F., Lu, H.-J., Zhu, Z.-Y., et al. (2017). Risk factors and preventions of breast cancer. *International Journal of Biological Sciences, 13*(11), 1387.

Szot, C. S., Buchanan, C. F., Freeman, J. W., & Rylander, M. N. (2011). 3D in vitro bioengineered tumors based on collagen I hydrogels. *Biomaterials, 32*(31), 7905–7912.

Szot, C. S., Buchanan, C. F., Freeman, J. W., & Rylander, M. N. (2013). In vitro angiogenesis induced by tumor-endothelial cell co-culture in bilayered, collagen I hydrogel bioengineered tumors. *Tissue Engineering Part C: Methods, 19*(11), 864–874. https://doi.org/10.1089/ten.tec.2012.0684.

Taubenberger, A. V., Bray, L. J., Haller, B., Shaposhnykov, A., Binner, M., Freudenberg, U., et al. (2016). 3D extracellular matrix interactions modulate tumour cell growth, invasion and angiogenesis in engineered tumour microenvironments. *Acta Biomaterialia, 36*, 73–85.

Thomas, E., Zeps, N., Rigby, P., & Hartmann, P. (2011). Reactive oxygen species initiate luminal but not basal cell death in cultured human mammary alveolar structures: A potential regulator of involution. *Cell Death & Disease, 2*(8), e189.

Trowell, O. A. (1959). The culture of mature organs in a synthetic medium. *Experimental Cell Research, 16*(1), 118–147.

Tung, Y.-C., Hsiao, A. Y., Allen, S. G., Torisawa, Y.-S., Ho, M., & Takayama, S. (2011). High-throughput 3D spheroid culture and drug testing using a 384 hanging drop array. *Analyst, 136*(3), 473–478.

Tyan, S.-W., Kuo, W.-H., Huang, C.-K., Pan, C.-C., Shew, J.-Y., Chang, K.-J., et al. (2011). Breast cancer cells induce cancer-associated fibroblasts to secrete hepatocyte growth factor to enhance breast tumorigenesis. *PLoS One, 6*(1), e15313.

van der Kuip, H. M. T., Sonnenberg, M., McClellan, M., Gutzeit, S., Gerteis, A., Simon, W., et al. (2006). Short term culture of breast cancer tissues to study the activity of the anticancer drug taxol in an intact tumor environment. *BMC Cancer, 6*(1), 1–11.

Vandamme, T. F. (2014). Use of rodents as models of human diseases. *Journal of Pharmacy & Bioallied Sciences, 6*(1), 2.

Viale, G. (2012). The current state of breast cancer classification. *Annals of Oncology, 23*, x207–x210.

Wahba, H. A., & El-Hadaad, H. A. (2015). Current approaches in treatment of triple-negative breast cancer. *Cancer Biology & Medicine, 12*(2), 106.

Wang, Y., Mirza, S., Wu, S., Zeng, J., Shi, W., Band, H., et al. (2018). 3D hydrogel breast cancer models for studying the effects of hypoxia on epithelial to mesenchymal transition. *Oncotarget, 9*(63), 32191.

Wang, S. E., Narasanna, A., Perez-Torres, M., Xiang, B., Wu, F. Y., Yang, S., et al. (2006). HER2 kinase domain mutation results in constitutive phosphorylation and activation of HER2 and EGFR and resistance to EGFR tyrosine kinase inhibitors. *Cancer Cell, 10*(1), 25–38.

Wang, X., Sun, L., Maffini, M. V., Soto, A., Sonnenschein, C., & Kaplan, D. L. (2010). A complex 3D human tissue culture system based on mammary stromal cells and silk scaffolds for modeling breast morphogenesis and function. *Biomaterials, 31*(14), 3920–3929.

Williams, A. D., Payne, K. K., Posey, A. D., Hill, C., Conejo-Garcia, J., June, C. H., et al. (2017). Immunotherapy for breast cancer: Current and future strategies. *Current Surgery Reports, 5*(12), 1–11.

Wind, N., & Holen, I. (2011). Multidrug resistance in breast cancer: From in vitro models to clinical studies. *International Journal of Breast Cancer, 2011*, 967419.

Wronski, A., Arendt, L., & Kuperwasser, C. (2015). Humanization of the mouse mammary gland. In *Mammary stem cells* (pp. 173–186). Springer.

Zhong, A., Wang, G., Yang, J., Xu, Q., Yuan, Q., Yang, Y., et al. (2014). Stromal–epithelial cell interactions and alteration of branching morphogenesis in macromastic mammary glands. *Journal of Cellular and Molecular Medicine, 18*(7), 1257–1266.

CHAPTER FIVE

Recapitulating human skeletal muscle in vitro

Anna Urciuolo[a,b], Maria Easler[a], and Nicola Elvassore[c,d,e,]*
[a]Institute of Pediatric Research (IRP), Fondazione Città della Speranza, Padova, Italy
[b]Department of Molecular Medicine, University of Padua, Padova, Italy
[c]Department of Industrial Engineering, University of Padua, Padova, Italy
[d]Veneto Institute of Molecular Medicine, Padova, Italy
[e]University College London Great Ormond Street Institute of Child Health, London, United Kingdom
*Corresponding author: e-mail address: n.elvassore@ucl.ac.uk

Contents

1. Lessons from skeletal muscle anatomy and physiology   183
2. Lesson from skeletal muscle development   185
3. Lesson from skeletal muscle regeneration   187
4. Emerging approaches for human skeletal muscle in vitro models   191
    4.1 Multicellular composition   191
    4.2 3D organization and mechanical properties   193
    4.3 Mechanical and electrical stimulations   196
5. Conclusions and future perspectives   198
Acknowledgments   200
References   200

Despite the tremendous impact of animal models in underpinning the role of specific players involved in myogenesis and in pathogenesis of muscle disorders (Bushby, Collins, & Hicks, 2014; Guiraud et al., 2017), species-specific responses have limited the study of human skeletal muscle tissue (Hu, Charles, Akay, Hutchinson, & Blemker, 2017; Slater, 2017). Therefore, there is an urgent need of human models, which can help elucidate human myogenesis and patient-specific pathophysiologically mechanisms for the development of novel therapeutical strategies. Ideally a human skeletal muscle model should recapitulate the complex environment and

Advances in Stem Cells and their Niches, Volume 6
ISSN 2468-5097
https://doi.org/10.1016/bs.asn.2021.10.003

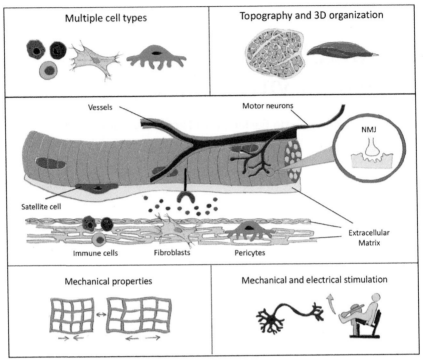

Fig. 1 Parameters necessary for recapitulating skeletal muscle tissue in vitro. Central panel represents close interactions between the myofibers, the ECM and supportive cell populations indispensable for tissue homeostasis in vivo. The top and bottom panels represent a number of properties which should be considered for in vitro modeling of the skeletal muscle tissue such as: multicellular composition, anisotropic alignment and 3D organization, elasticity and both mechanical and electrical stimulation.

three-dimensional (3D) organization of the skeletal muscle including multiple cell types present within the tissue in vivo and their spatiotemporal interactions (Fig. 1). Starting from the basic knowledge of muscle anatomy and physiology and learning from dynamic events such as development and regeneration of the skeletal muscle, in recent years, scientists have been able to apply bioengineering technologies to generate novel and complex human skeletal muscle models that better recapitulate the native tissue complexity, allowing the investigation of pathophysiological relevant phenotypes (Table 1).

Table 1 Published literature review between years 2015 and 2021 on the topic of emerging experimental approaches for human skeletal muscle models.

Experimental approaches	Experimental set-up	Applications	References
	Isogenic hiPSC–derived MC, EC, PC and MN assembled into a 3D artificial muscle	Laminopathies modeling	Maffioletti et al. (2018)
Multicellular composition	hiPSC–derived 3D myobundles and patient hiPSC–derived MNs co-culture	ALS modeling and evaluation of MN degeneration after drugs treatment	Osaki et al. (2018)
	hiPSC–derived 3D myobundles or 2D cell culture	Laminopathies modeling and comparison of 2D/3D environment on model outcome	Steele-Stallard et al. (2018)
	3D human primary myogenic constructs and hESC–derived MN clusters co-culture	MG modeling via MG patient serum treatment	Bakooshli et al. (2019)
	hiPSC–derived neuromuscular organoids	Human NMJ characterization and formation	Martins et al. (2020)
3D organization: Biomaterials and techniques	Micropatterned photosensitive hydrogels with physiological stiffness to culture from primary myoblasts	Healthy and DMD tissue–on the-chip constructs	Serena et al. (2016)
	Electrospun PM nanofibers coated with either laminin or collagen for 3D primary human myoblast culture	Investigation of myoblast proliferation, migration and differentiation	Zahari, Idrus, and Chowdhury (2017)
	3D bioprinting using integrated system of durable PCL support hydrogel, sacrificial gelatin and hMPC-laden fibrin hydrogel	Generation of transplantable printed tissue/organ constructs	Kim et al. (2018)

Continued

Table 1 Published literature review between years 2015 and 2021 on the topic of emerging experimental approaches for human skeletal muscle models.—cont'd

Experimental approaches	Experimental set-up	Applications	References
	2-photon sensitive hydrogel 3D bioprinting of murine muscle stem cell–derived myogenic cells and fibroblasts	3D constructs for in vitro modeling and spatially controlled in vivo cell delivery	Urciuolo et al. (2020)
	Decellularized murine diaphragm recellularized with pediatric hMPCs	Role of native ECM in modeling myogenesis	Trevisan et al. (2019) and Boso et al. (2021)
	Decellularized porcine diaphragm-derived hydrogels for 3D co-culture of hMPCs and fibroblasts	Transplantable repopulated or acellular scaffold development to support myogenic cell repopulation and muscle defect repair	
Dynamic cultures	Pillar-suspended myoblast-seeded GelMA microfibers cultured under uniaxial strain conditions of various tension strength	Investigation of alignment, migration and differentiation of myogenic cells	Chen et al. (2020), Heher et al. (2015), Michielin, Serena, Pavan, and Elvassore (2015), and Khodabukus et al. (2019)
	Myoblast-laden fibrin rings are suspended under static conditions with magnetically controlled bioreactor	Enhanced differentiation and maturation of myogenic cells	
	Microfluidic-based stretching device to assess healthy donor and DMD myoblast and myotube behavior	Modeling DMD and membrane permeability under static and dynamic conditions	
	Electrical stimulation was applied to 3D fibrin-based hydrogel construct seeded with human primary myogenic cells	Amelioration of human myogenic cell differentiation and maturation as a result of electrical stimulation	

Abbreviations: MN (motor neuron); ALS (amyotrophic lateral sclerosis); hiPSC (human induced pluripotent stem cells); hESC (human embryonic stem cells); MG (myasthenia gravis); NMJ (neuromuscular junction); MC (myogenic cells); EC (endothelial cells); PC (pericytes); DMD (Duchenne muscular dystrophy); PM (polymethacrylate); PCL (poly(ε-caprolactone)); hMPC (human myogenic progenitor cells); GelMA (gelatin methacrylate).

1. Lessons from skeletal muscle anatomy and physiology

The basic cellular unit of skeletal muscle is represented by the myofiber, an elongated and multinucleated postmitotic cell able to contract (Exeter & Connell, 2010). A parallel structural 3D organization of groups of myofibers is required to allow force generation upon myofiber contraction (Mukund & Subramaniam, 2020). This 3D anatomical organization of myofibers is guaranteed by the presence of the extracellular matrix (ECM), surrounding each single muscle fiber and at group level of organization. The whole muscle is wrapped by the collagen fibrils-enriched epimysium, while perimysium hold groups of myofibers together ensuring proper force transmission (Halper & Kjaer, 2014). Perimysium, marked by several layers of transversely oriented collagen networks, is highly permeated by numerous blood vessels and nerves (Latroche et al., 2015). Each myofiber in turn is enveloped by the endomysium supplied by nerve endings and extensive networks of capillaries running along the length of the myofiber with collagen fibers exhibiting random organization to avoid interference with contraction (Ahmad, Shaikh, Ahmad, Lee, & Choi, 2020; Latroche et al., 2015) and a specialized basement membrane. The basement membrane of the myofiber is a bipartite structure made of reticular and basal laminas (Garg & Boppart, 2016). The reticular lamina neighbors the endomysium and contains proteoglycan-rich gel infused with fibronectin and collagen fibrils, and the basal lamina is proximal to sarcolemma and is composed mostly of laminin and collagen IV and VI (Ry et al., 2017). Stabilization of the interaction between myofiber's sarcolemma and the ECM is achieved by the interaction of specific receptors that through dystrophin-glycoprotein complex (DGC) secure the binding of the extracellular environment to the cytoskeletal components of the myofiber, preventing damage of the sarcolemma during contraction (Csapo, Gumpenberger, & Wessner, 2020). The importance of ECM and ECM-binding proteins for normal skeletal muscle development, function and regeneration is evident from loss-of-function animal models characterized by lethal outcome either during embryonic development or shortly after birth (Yao, 2017) or by their muscular phenotype, which often model human muscular disorders (Ahmad, Shaikh, Ahmad, Lee, & Choi, 2020; Cescon et al., 2018; van Ry et al., 2017).

The development, homeostasis and function of skeletal muscles are not only guaranteed by a precision in structural three-dimensional organization of the myofibers, but also by an elaborate interaction of the muscle fibers with other cell types such as those of nervous, vascular and mesenchymal origin (Chapman, Meza, & Lieber, 2016; Huey, 2018; Witzemann, 2006). In particular, the distinguishing feature of the skeletal muscle from other types, such as cardiac and smooth muscles, is the ability to voluntarily control the muscle contraction. This voluntary action relies on the establishment and function of the neuromuscular junction (NMJ) during development. The NMJ represents the connection point between the motor neuron (MN) axons of the nervous system and the myofibers of the skeletal muscles. A functional NMJ is characterized by a tripartite assembly where the presence of terminal Schwann cells (tSC) ensures proper development, maintenance and remodeling of the NMJ (Alhindi et al., 2021). The MN cell bodies are located within the ventral horn of the spinal cord and their axonal projections, organized in bundles that give rise to the nerve, extend to reach their respective skeletal muscle targets sometimes located at a great distance from the MN cell body (Stifani, 2014). Each MN can maintain connections with multiple myofibers thus organizing a motor unit (Alvarez-Suarez, Gawor, & Prószyński, 2020). The ability of the skeletal muscle to induce voluntary contractions depends on the excitatory properties of the MN and the ability of the myofibers to respond to the chemical stimulus received from the nerve terminal. Briefly, when action potential reaches the MN nerve terminal, it triggers the release of the neurotransmitter acetylcholine (Ach) into the synaptic cleft, the junctional area between the presynaptic compartment (the MN terminal) and the postsynaptic compartment on the myofibers (the motor end plate). Following neurotransmitter binding of the Ach receptors (AChR) of the motor end plate, resulting end-plate potential initiates depolarization within the muscle fiber, subsequent release of the calcium stores from the sarcoplasmic reticulum and, finally, the initiation of contraction (Exeter & Connell, 2010; Heikkinen, Härönen, Norman, & Pihlajaniemi, 2020). The first communication between the nerves and the muscles takes place early during development which sets the stage for future function and homeostasis of the skeletal muscle tissue as a whole. In fact, a number of myopathies originate from disruption of the crosstalk between the myofibers and the nerves (Gromova & La Spada, 2020). Additionally, maintaining this communication is necessary for the homeostasis of the skeletal muscle. This is demonstrated by ailments such as spinal muscular atrophy and amyotrophic lateral sclerosis where the loss of MNs has indirect deteriorating effect on the skeletal muscle tissue resulting in fatigue, atrophy and, later, inability to carry out daily vital functions such as walking, eating and breathing (Bonaldo & Sandri, 2013; Messina & Sframeli, 2020; Takeda, Kitagawa, & Arai, 2020).

2. Lesson from skeletal muscle development

Skeletal muscle development, studied mostly in model organisms such as frog, zebrafish, chick and mouse, relies on the combination of cell autonomous pathways and cell-cell and cell-environmental communications defined via morphogens and signaling pathways in a spatiotemporal fashion (Deries & Thorsteinsdóttir, 2016). The complex and finely tuned events happening during myogenesis also involve interactions with the surrounding mesenchymal masses, developing nervous tissue and long-range signals from target locations throughout the body. The identification of key molecular and cellular players involved in myogenesis has recently been used from scientists to trigger myogenesis in vitro guiding pluripotent stem cells (PSCs) or other stem cells toward skeletal muscle tissue identity, allowing the generation of promising human muscle models. Here, we summarized the most relevant players acting during skeletal muscle development.

Apart from neck muscles, skeletal muscles in vertebrates originate from paraxial mesoderm, that is laid down at the time of gastrulation flanking both sides of the notochord and neural tube (Chal & Pourquié, 2017). Paraxial mesoderm later undergoes segmentation into somites in the process of somitogenesis where the external morphogenic cues guide compartmentalization of somites into three regions: sclerotome, syndetome, dermomyotome. The sclerotome and syndetome give rise to axial bones, tendons and connective tissue, and the dermomyotome contains proliferating progenitors of the skeletal muscle, brown fat, endothelial cells and dorsal dermis (Deries & Thorsteinsdóttir, 2016). During early development, two waves of myogenesis takes place: primary and secondary (Chal & Pourquié, 2017). Primary myogenesis (in mice E8.75–E11) refers to the first wave of myogenic differentiation during embryonic stages of development in vertebrates (Biressi et al., 2007). Primary myoblasts, originating from dorsomedial lip of dermomyotome and expressing the key myogenic transcription factor myogenic factor 5 (Myf 5), delaminate from dermomyotome and differentiate into mononucleated myocytes establishing primary myotome. Initially, only a part of embryonic myoblasts differentiates into myocytes and first populate the somite region followed by their migration into the trunk and limbs establishing the primary myofibers (Biressi et al., 2007). The migratory path of this myogenic population can be observed by reporter animal studies that follow the expression of paired box 3 (Pax3) transcription factor characterizing the location of the primary myotome (Hurren et al., 2015). Secondary myogenesis (in mice E14.5–E17.5) involves differentiation of fetal myoblasts which downregulate the expression of Pax3 transcription factor and upregulate the expression of paired box 7 (Pax7) transcription factor. The migration of secondary myogenic progenitors, their differentiation to committed myoblasts (expressing the myoblast determination protein 1 (MyoD) and fusion with each other or primary fibers contributes to the expansion of newly established muscle in the developing fetus (Yin, Price, & Rudnicki, 2013). Moreover, a group of Pax7-positive progenitors establish a population of quiescent satellite cells (SCs), which settle between the myofibers and their basal lamina (Yin et al., 2013).

The molecular triggers of somitogenesis include morphogens released from the surrounding developing tissues, including ectoderm (Wnt signaling), notochord (Shh), neural tube (Shh from the floor plate) and lateral plate mesoderm (BMP) (Deries & Thorsteinsdóttir, 2016). Moreover, during early myogenesis, the intimate relationship between the developing myotome and the surrounding connective tissue is evident. For example, Tbx5 expression in muscle connective tissue (MCT) had been shown to play an important role in muscle bundle morphogenesis during embryonic development and its ablation induced limb deformations (Besse et al., 2020). Similarly, the presence of MCT fibroblasts expressing Tcf4 was found to be necessary for precise muscle morphogenesis and for appropriate muscle type switch during development (Mathew et al., 2011). In addition, the inhibitory and activating cues from surrounding developing tissues regulate the dynamics of myogenesis (Amthor, Christ, Weil, & Patel, 1998; Costamagna, Mommaerts, Sampaolesi, & Tylzanowski, 2016).

Another known regulator of myogenesis, with great impact on maturation of myofibers during development, is innervation. It has been demonstrated that NMJ formation firstly happens in primary myofibers. Indeed, the innervation site of the primary myofibers is the region around which the secondary myofibers aggregate to fuse into an expanded anlage (Duxson, Usson, & Harris, 1989). The formation of the NMJ requires complex interactions between the tri-partite NMJ components. To commence the interaction, the myofibers display pre-patterned distribution of AchR clusters even before the initiation of direct communication with MN axon terminal (Yang et al., 2001). Upon the arrival of the axon terminal, bidirectional communication between the neuron and the myofiber, stabilizes the AchR clusters. One of the major mechanisms implicated in AchR cluster stabilization is Agrin-Lrp4-MuSK pathway where MN terminal-derived agrin binds its postsynaptic receptor Lrp4, which induces phosphorylation of muscle-specific tyrosine kinase initiating production of Rapsyn protein necessary for AchR stabilization (reviewed by Wu, Xiong, & Mei, 2010). Although each individual myofiber will become innervated by multiple MNs during development, to establish an adequate functional motor unit each myofiber has to be innervated by a single MN. This phenomenon is possible due to axon pruning which takes place after birth (in mice). It has been demonstrated that, in addition to guiding synaptic maturation, tSCs play a role in actively eliminating multiple axonal sprouts from the NMJ to enable final status of myofiber as a monoinnervated construct. While muscle fibers express nicotinic AchR (nAchR), tSCs express muscarinic AchR (mAchR) as well as purinergic receptors (P2Y1R) capable of binding nerve-derived Ach and adenosine (hydrolyzed ATP contained within Ach vesicles) respectively. Here, asynchronous firing of MN axons at birth, guides the detection by tSC of stochastically "stronger" candidate for NMJ formation. In turn, Ach and adenosine signaling in tSC induces it to secrete even more ATP into extracellular space, thus, increasing the level of purinergic signaling within the neuron and, finally, potentiating innervation (Alvarez-Suarez, Gawor, & Prószyński, 2020). On the other hand, there is evidence that tSCs play an active role in countering polyinnervation through physical and phagocytic methods in order to displace a competing axon by extending finger-like protrusions between the terminal and the myofiber or by extending phagocytic processes into the receding axons (Smith, Mikesh, Lee, & Thompson, 2013).

3. Lesson from skeletal muscle regeneration

Skeletal muscle tissue possesses a remarkable ability for regeneration during the lifetime of the organism. The knowledge derived from studies on skeletal muscle regeneration has opened new perspectives for the definition of cellular, molecular and mechanical properties that should be considered during skeletal muscle tissue model development. Here we introduce the major players involved during skeletal muscle regeneration that have been eventually applied for modeling skeletal muscle in vitro.

The major cell contributor of newly formed myofibers during homeostasis and after injury of adult skeletal muscle is the pool of SCs (Sambasivan et al., 2011; Tedesco, Dellavalle, Diaz-manera, Messina, & Cossu, 2010; Messina et al., 2010, Yin et al., 2013). The preservation of SC stemness is guaranteed by their *niche*, the stem cell environment providing guiding cues for quiescence, activation, proliferation and differentiation of the SC (Yin et al., 2013). After skeletal muscle injury, inflammatory effectors such as neutrophils and macrophages are recruited to clear cellular debris (Yang & Hu, 2018). At the same time, the activation of the SC takes place with endogenous and exogenous stimuli coordinating the event. The cytokines (TNF-α, IFN-γ and IL-1) secreted by neutrophils and pro-inflammatory macrophages (M1), attract even more myeloid cells to the site of injury and reinforce the activation of the SCs (Forcina, Cosentino, & Musarò, 2020). However, timely involvement of the anti-inflammatory macrophages (M2) has been found absolutely necessary for appropriate SC differentiation progression and ECM remodeling (Tidball & Wehling-Henricks, 2007; Arnold et al., 2007). As well described in detailed reviews, upon activation, SCs undergo proliferation, self-renewal and differentiation to reconstitute the SC pool and give rise to myogenic cells able to fuse together or fuse to damaged fibers for the regeneration of myofibers (Tedesco et al., 2010; Yin et al., 2013; Sousa-Victor, García-Prat, & Muñoz-Cánoves, 2021). Moreover, during this phase of regeneration, the activity of connective fibroblasts and ECM remodeling can also be referred to as an essential stage required during muscle regeneration (Murphy, Lawson, Mathew, Hutcheson, & Kardon, 2011; Urciuolo et al., 2013). Muscle fibroblasts reestablish the ECM for sustaining SC stemness, and for overall structural organization of the regenerating tissue. Consequently, various ECM components are deposited around newly formed myotubes and myofibers, a process finely orchestrated by cues emanating from anti-inflammatory macrophages during this time. In fact, it has been demonstrated in mouse models that in the absence of M2 macrophages during muscle regeneration, differentiation and maturation of the myofibers is delayed in addition to deregulated structural distribution and organization of the collagen fibers (Garg, Corona, & Walters, 2015; Gillies et al., 2017). The last stage of regeneration involves maturation and functional recovery of the reconstructed myofibers including angiogenesis and innervation.

In vivo myogenesis and angiogenesis happen concomitantly during regeneration and it is clear from in vitro studies that both myoblasts and angioblasts influence each other's fate during differentiation through paracrine signaling (Latroche et al., 2017). Instead, innervation of the regenerating myofiber relies on molecular cues retained by the basal lamina of the NMJ that seem to guide the nerve terminal of the surviving MN axon to a newly established myofiber (Slater & Schiaffino, 2008).

As mentioned earlier, SCs are indispensable for skeletal muscle regeneration, and the balance between self-renewal and differentiation has tremendous impact on the regenerative outcome. It is known that overall proliferative and differentiation potential of the SC is tightly regulated by the *niche* environment in which they reside. In fact, the balancing act between symmetric and asymmetric cell divisions is carefully tuned to power the yield of both the committed myogenic progenitors and the pool of Pax7-positive quiescent SCs depending on the tissue needs (Yin et al., 2013). This balance is maintained by a variety of signaling pathways often imposed by cell–cell and cell–ECM interactions. One of the cell–cell interactions involves Notch signaling pathway. The close association of Notch-3 receptor expressed on SC with Notch ligand Delta1-expressing myofibers is necessary to maintain Notch signaling which is required for maintenance of quiescent status of the SCs. The ablation of Rbpj (downstream effector of Notch signaling) in SCs in developing mice resulted in premature SC differentiation and loss of stem cell pool undermining future regeneration (Philippos et al., 2012). Recent reports also demonstrate that Delta-like-1 (Dll1) Notch ligand oscillatory expression during SC activation regulate the fate of the daughter cells during asymmetric cell division ensuring the sustainability of the stem cell pool (Zhang et al., 2021). However, inactivation of Notch signaling in myoblasts is required for their fusion. This is done through asymmetric partitioning of Numb during cell division. Following asymmetric division, in the committed daughter cell Numb activates expression of Desmin and ubiquitinates NICD, pushing the cell toward differentiation (Conboy & Rando, 2002). Another major regulator, Wnt signaling pathway, had been indicated as an important modulator of myogenic progenitor differentiation, with Notch and Wnt pathways demonstrating antagonistic relationship in terms of driving the commitment to differentiation. In fact, timely coordination between the two pathways seems to be necessary during early and late stages of regeneration to ensure both the maintenance of the stem cell pool and formation of newly committed progenitors (Brack, Conboy, Conboy, Shen, & Rando, 2008; von Maltzahn, Florian Bentzinger, & Rudnicki, 2012). Other indispensable factors for activation of SC proliferation and differentiation are presented by the ECM components of the SC niche (Gattazzo, Urciuolo, & Bonaldo, 2014). One of such prominent activators appears to be the hepatocyte growth factor (HGF). In homeostatic conditions, HGF is found within the SC niche sequestered by heparan sulfate proteoglycans (HSPG) of the basal lamina. During regeneration, both nitric oxide synthase (NOS) and nitric oxide (NO) guide the behavior of myogenic and macrophage cell populations. Specifically, it is found that the increase in NO leads to activation of matrix metalloproteases (MMPs) releasing HGF from HSPGs which binds to c-Met receptor on the surface of quiescent SC, induce cell cycle progression and differentiation (Allen, Sheehan, Taylor, Kendall, & Rice, 1995; Buono et al., 2012; Rigamonti et al., 2013; Tatsumi, Anderson, Nevoret, Halevy, & Allen, 1998; Yamada et al., 2006). While HGF functions during the initial stages of regeneration, various isoforms of fibroblast growth factor (FGF) and insulin-like growth factor (IGF) act in consortium to guide myogenic differentiation, fusion and maturation of differentiating myoblasts (Neuhaus et al., 2003).

In addition, ECM components have been shown to influence SC division. Specifically, the proteins fibronectin (Bentzinger et al., 2013), collagen VI (Urciuolo et al., 2013) and collagen V (Baghdadi et al., 2018) as well as the proteoglycans syndecan 3, syndecan 4, perlecan, and decorin (Brack et al., 2008; Cornelison et al., 2004) have been identified as the niche constituents influencing the balance between differentiation and self-renewal and, thus, the maintenance of skeletal muscles' regenerative capacity.

The past 20 years have witnessed ever-growing evidence that the mechanical properties of biological tissues are fundamental for cellular behavior and consequent tissue functionality (Guimarães, Gasperini, Marques, & Reis, 2020). With each contraction–relaxation cycle, a SC is expected to experience tensile, shear, and compressive stresses, and through cell–extracellular matrix interactions, also gauge the stiffness of the niche. Via mechanoreceptors, cells can sense these biophysical parameters of the niche, which serve to physically induce conformational changes that impact biomolecule activity, and thereby alter downstream signal transduction pathways and ultimately cell fate (Li, McKee-Muir, & Gilbert, 2018). In particular, the elastic modulus of skeletal muscle, often expressed by the

Young's modulus, has been shown to be a key mechanical property able to influence muscle regeneration and SC stemness maintenance (Li et al., 2018). Previous measurements using atomic force microscopy and shear rheometry demonstrated that the elastic modulus of the healthy skeletal muscle is around 12 kPa (Engler et al., 2004; Gilbert et al., 2010). In vitro studies exploring culture of primary SC on substrates of different stiffness demonstrated that substrates presenting native-like environment in terms of stiffness for SC culture allow for SC-self renewal (Monge et al., 2017; Gilbert et al., 2010; A. Urciuolo et al., 2013) and can promote maintenance of their quiescent state (Quarta et al., 2016). In another study, the hydrogel substrate of stiffness of about 15 kPa induced efficient differentiation of human myogenic progenitors when compared to stiffer scaffolds (Serena et al., 2010). Alteration of skeletal muscle stiffness and ECM deposition can be observed in aging and disease muscles, negatively influencing the muscle regenerative capacity (Csapo et al., 2020; Stearns-Reider et al., 2017).

4. Emerging approaches for human skeletal muscle in vitro models

Based on what was discussed previously, the emerging approaches to generate human skeletal muscle in vitro models are mainly based on the reproduction of the following features of the tissue: (1) multicellular composition; (2) 3D organization and mechanical properties; (3) mechanical and electrical stimulation. In particular, recent advancements in stem cell biology, biomaterial and tissue engineering technologies made it possible to better mimic the in vivo features of skeletal muscle tissue in vitro.

4.1 Multicellular composition

The basic cell type required for the generation of human skeletal muscle model is represented by cells that possess myogenic potency. The traditional cell source used for human skeletal muscle modeling use primary myoblasts derived from healthy donors or patient muscle biopsies. The advantage of using primary myoblast culture is related to their proper myogenic activity and the preservation of patient-specific heterogeneity. However, a relevant obstacle in primary myoblasts use is related to their limited expansion in culture (Tabakov, Zinov'eva, Voskresenskaya, & Skoblov, 2018). On the contrary, the use of commercial myogenic cell lines allows for multiple passages of cell culture reducing the onset of senescence. Nevertheless, despite the convenience of predictable and efficient outcome of myoblast differentiation, commercially available myoblast cell lines are marked by either immortalized or transformed phenotype which does not accurately reflect transcriptional profile or myoblast cell behavior in vivo (Kaur & Dufour, 2012; Pan, Kumar, Bohl, Klingmueller, & Mann, 2009; Wilson, Breen, Lord, & Sapey, 2018). In the recent years, additional expandable source of myogenic cells came into play through the availability and manip-ulation of human pluripotent stem cells (hPSC). Both human embryonic stem cells (hESC) and human induced pluripotent stem cells (hiPSC) are characterized by a remarkable ability to self-renew, thus representing an excellent expandable cell source for the differentiation of patient-specific myogenic cells.

As previously discussed, skeletal muscle is not only composed by myofibers, but other cell types play an essential role in skeletal muscle tissue homeostasis, growth and function, including MNs, endothelial cells (ECs), smooth muscle cells (SMCs), pericytes, fibroblasts and fibroadipogenic pro-genitors, immune cells, and other interstitial cells (Mukund & Subramaniam, 2020; Yin et al., 2013). Based on this, the emerging approaches to model skeletal muscle tissue attempt to combine multiple key cell types.

Thanks to the co-culture of myogenic cells with patient iPSC-derived MNs, Osaki and colleagues generated a model of human skeletal muscle where NMJs were formed and it was possible to model amyotrophic lateral sclerosis (ALS) (Osaki, Uzel, & Kamm, 2018). MN degeneration and reduced muscle contraction pathological features were recovered with application of ALS drug candidates (Osaki et al., 2018). Autoimmune disease myasthenia gravis was instead modeled thanks to the use of hPSC-derived MN co-culture with myogenic cells by adding patient serum to healthy 3D muscle models and quantifying muscle force (Bakooshli et al., 2019; Martins et al., 2020; Vila, Yihuai, & Vunjak-Novakovic, 2020). Despite these relevant results, the maturation of NMJ requires multiple cell types apart from MNs and myogenic cells. In the recent study by Martins and colleagues and colleagues, self-assembled neuromuscular organoids where derived from isogenic multilineage differentiation of hPSCs (Martins et al., 2020). Interestingly, such strategy allowed the generation of functional, mature NMJs, containing also tSCs and interneural networks (Faustino Martins et al., 2020). Another strategy that attempts at creating multilineage isogenic 3D in vitro skeletal muscle model starting from hPSCs has seen the co-culture of vascular endothelial cells and pericytes individually differ-entiated from the same hiPSC cell line as the myogenic population

(Maffioletti et al., 2018). Starting from patient-derived hiPSCs, myogenic cells were used to mimic some feature of skeletal muscle laminopathies (Maffioletti et al., 2018; Steele-Stallard et al., 2018). It is worth to underline that such strategies are still under optimization and further investigation is required to reveal the potential use of these approaches for modeling congenital diseases and human myogenesis.

4.2 3D organization and mechanical properties

The use of biomaterials as 3D scaffolds where myogenic cells can be cultured represents one of the most used strategies to permit myotube alignment and to support their contraction, while eventually integrating multiple cell types. As discussed before, one of the prominent features of native skeletal muscle tissue is its sophisticated structural 3D organization that guides and maintains tissue homeostasis and function.

Hydrogel is the most prominent biomaterial used to create 3D skeletal muscle tissue. Hydrogel can be of natural or synthetic origin. Natural hydrogels are composed of one or more ECM components of either animal or plant origin (collagen, fibrin, gelatin, hyaluronic acid, agarose, alginate, silk) which by default can exert signaling cues reminiscent to those in vivo (Boso et al., 2020). This attractive bioregulatory feature, however, is weakened by relative instability of natural compounds and low degree of viscosity (Visscher et al., 2016). Alternatively, synthetic hydrogels are chemically fabricated materials (polyurethane, polyethylene glycol, polylactic acid, and polyvinyl alcohol) with tunable stiffness, stability and viscosity (Boso et al., 2020). Despite this advantage, synthetic hydrogels are inert in terms of biological guidance and more often than not ought to be ameliorated by either incorporation of naturally derived hydrogels or functionalization by signaling compounds to induce biologically relevant cellular responses. The major advantages of use of hydrogels for the generation of 3D skeletal muscle models is malleability of the dimensions, shape and mechanical properties of the cell-laden environment. The modification of elasticity and viscosity of the hydrogel depends on the degree of intermolecular crosslinking which can be classified as physical or chemical (Nguyen & West, 2002; Wang, Rivera-Bolanos, Jiang, & Ameer, 2019).

One of the challenges in modeling skeletal muscle is the formation of structurally competent constructs. In vivo skeletal muscle is characterized by its anisotropic alignment of myofibers tightly compacted into an aggregate of parallelly aligned bundles. To reinstate similar structural features during in vitro fabrication a variety of technologies had been explored to include micropattern fabrication, electrospinning and 3D bioprinting (Jana, Lan, & Levengood, and Miqin Zhang., 2016; Serena et al., 2016; Urciuolo et al., 2020; Zhao, Zeng, Nam, & Agarwal, 2009). Micropattern fabrication involves the use of patterned mask to imprint the desired structures onto the unpolymerized hydrogel, consequent polymerization and removal of the mask to yield the micropatterned polymerized hydrogel constructs ready for cell culture. Indeed, previous investigation of myogenic cell behavior in response to topographic microgrooves presented within the hydrogel revealed enhanced migration, attachment, differentiation and alignment of myogenic cells (Denes et al., 2019; Zhao et al., 2009). In addition to providing directional guiding, such micropatterned hydrogels were also found to support the contractility of the differentiating 3D skeletal muscle constructs which is another necessary feature of relevant 3D in vitro models (Cimetta et al., 2009). The electrospinning technique, on the other hand, provides a scaffold fabricated from hydrogel nanofibers with high surface to volume ratio and tunable parameters such as diameter, stiffness and orientation of nanofibers. Again, there is an increasing number of evidence that aligned nanofibers enhance directional migration and alignment of differentiating myoblasts (Xue, Tong, & Xia, 2018) simulating the migration of myoblasts along the "ghost" fibers during in vivo regeneration. Additional functionalization of the surface of the nanofibers allows for improved cell adhesion, differentiation and maturation (Gilbert-Honick et al., 2020; Narayanan et al., 2020; Zahari, Idrus, & Chowdhury, 2017).

Another widely used technique with huge perspectives for in vitro, in vivo and in situ applications, and aimed at spatially control 3D cell cultures, is 3D bioprinting. Rapid development of bioprinting technology in the last decade is associated with its numerous advantages. Bioprinting technology allows the generation of 3D cellular constructs thanks to the micrometric precision in layer-by-layer deposition of cell-laden bioink that can be converted from a liquid to a solid state. The identity of both the cell type and hydrogel-based bioink can be adjusted to closely mimic the native composition of the tissue. Indeed, a single experiment can enable multiplexing of different cell types and bioinks to reinstate the variation not only in the cell type composition but also in the environment in which the cells are found in vivo (Kang et al., 2016). Another advantage of 3D bioprinting is the tunability of mechanical properties of the printed bioink enabling regulation of cell behavior (Zhuang, An, Chua, & Tan, 2020). The spatial control of bioink deposition provides an opportunity to manage the tissue construct composition and scale. For example, simultaneous deposition of both permanent and sacrificial materials can permit development of larger constructs with valuable internal organization that allows for better nutrient access and distribution (Kim et al., 2018). This can be an important consideration in light of the decrease in the efficiency of large in vitro construct culture due to the presence of a necrotic core. Finally, 3D bioprinting process is highly reproducible enabling efficient standardization of the process and production of robust data.

Different modes of bioprinting are currently employed, among them are inkjet, extrusion, laser-assisted and stereolithographic bioprinting (Ostrovidov et al., 2019). Although the speed, cost and resolution of printing varies depending on the choice, majority of bioprinting techniques offer high cell viability (up to 95%) at the end of the process which expands applicability of such in vitro constructs to translation medicine (Costantini et al., 2017).

In addition to traditional 3D-bioprinting methods, the use of photo-crosslinkable hydrogels can expand the utility of the bioprinting methods to beyond in vitro culture. Recently developed 2-photon sensitive-hydrogel has shown to sustain myogenic cell viability and enhance their alignment and differentiation in both in vitro and in vivo settings (Urciuolo et al., 2020). The implementation of such hydrogel permits 4D experimental control allowing for change of hydrogel parameters such as stiffness and dimensions at a desired point of time.

Although the use of individual native, synthetic or hybrid hydrogels associated with fabrication technique had proven useful in the generation of 3D skeletal muscle models, none have come close in replicating the biochemical, structural and functional complexity of the native tissue.

In recent years, decellularized tissues and organs have become a tool that overcome such limit providing native-like scaffolds (Urciuolo & De Coppi, 2018). The decellularization process enable the removal of nuclear and cellular components of the tissue-resident cells, but preserves ECM composition, structure and biomechanical properties of the native tissue (Anna Urciuolo & De Coppi, 2018). By preserving the structural integrity of the native ECM and acting as a reservoir of growth factors and small molecules, decellularized tissues and organs provide guidance cues for cell migration, alignment and homing, reestablishing native topography and configuration of cell components (Gresham, Bahney, & Kent Leach, 2021). Apart from their application in tissue engineering and regenerative medicine approaches, decellularized muscles have been used in vitro to reproduce skeletal muscle in vitro models.

Stearns-Reider and colleagues used decellularized skeletal muscle derived from both young and old mice and repopulated with human muscle stem cells to mimic aging and investigate the mechanotransductive pathways of fibroblast matrix deposition (Stearns-Reider et al., 2017).

Decellularized skeletal muscle has also been recellularized to create models in which myogenesis and regeneration can be investigated. Trevisan and colleagues decellularized mouse diaphragm muscle and repopulated with human muscle precursor cells in an effort to create a humanized diaphragm model (Trevisan et al., 2019). The construct was shown to regenerate after exposure to cardiotoxin (snake venom used to damage muscle fibers and initiate regenerative tissue responses), as well as maintain its stem cell niche (Trevisan et al., 2019). It has recently been shown that decellularized skeletal muscle scaffolds retain their neurotrophic ability by studying the directionality and the rate of axonal growth in a 3D in vitro construct (Raffa et al., 2020). This discovery provides a valuable tool for studying the role of the skeletal muscle ECM in regeneration.

Finally, due to its unique tissue-specific composition and important signals for cell fate determination, decellularized muscle have also been recently used to generate biomimetic bioinks and hydrogels that can be used to produce customized scaffolds with different biofabrication approaches (Baiguera et al., 2020; Boso et al., 2021; Giobbe et al., 2019; McCrary, Bousalis, Mobini, Song, & Schmidt, 2020).

4.3 Mechanical and electrical stimulations

As discussed above, skeletal muscle in vivo is subject to environmental dynamic stimuli, such as mechanical deformation and electric stimulations. Therefore, with the advent of 3D skeletal muscle models, such stimulations have been applied thanks to the use of bioreactors with the aim to improve alignment, maturation and functionality of the myofibers as well as to provide a surrogate neural input for muscle contraction (Kasper, Turner, Martin, & Sharples, 2018).

Mechanical load is a necessary component during skeletal muscle development and homeostasis. Indeed, during development, continuous limb elongation during rapid developmental growth imposes static stretching stimulus which aids in differentiation and maturation of myogenic progenitors (Heher et al., 2015). Whereas dynamic mechanical load, such as associated with movement in postnatal stage of development, serves to maintain contractile protein expression (Heher et al., 2015). In vitro studies employing tendon-like posts between which the muscle constructs are organized produce passive static strain which aids in myoblast differentiation, fusion and alignment producing myotubes of larger diameter, length and enhanced organization of sarcomeric proteins (Heher et al., 2015). Dynamic strain can be introduced by employing bioreactors, devices allowing for dynamic cell culture where degree and frequency of displacement can be manipulated to investigate the effect of mechanical stimulus on myogenic cell behavior. Indeed, it was found that application of mechanical stimulation through dynamic myogenic cell culture induces an increase in the expression of insulin growth factor-1 and myosin genes, resulting in enhanced maturation profile of the myotubes and increase in their diameter compared to myotubes cultured without mechanical stimulation (Cheema et al., 2005; Chen et al., 2020; Powell, Smiley, Mills, & Vandenburgh, 2002). Another application of dynamic in vitro modeling is to investigate and compare the response of healthy and diseased in vitro cultures. For example, Michielin and colleagues used microfluidic-based mechanical stimulation to compare the effect of the degree and frequency of the strain on healthy or dystrophic cell culture which revealed increased membrane permeability of the diseased constructs in response to mechanical stimulation (Michielin et al., 2015). Thus, the tunability of the in vitro construct, in which a number of stimuli can be manipulated, provides a wide range of experimental agendas that can be addressed relating to modeling of the development, pathophysiology, regeneration and the homeostasis of the skeletal muscle.

In vivo, neural stimulation assures maintenance of myofiber functionality and identity. In vitro, in the absence of neural component in cell culture, electrical stimulation can be utilized to investigate the influence of MN effect on myofiber function and maturation or mimic exercise conditions. It is known that the frequency of MN firing affects transcriptional profile of the myofiber type differing between slow and fast phenotype where slow twitch profile is innervated by "slow" MNs with firing frequencies of about 10–20 Hz while "fast" MNs are characterized by short bouts of high frequency firing (~100 Hz) (Cetin, Beeson, Vincent, & Webster, 2020). Previous transposition experiments, where fast and slow phenotypes of the MN and the myofiber were mismatched, confirmed mutual relationships between the neural input and the muscle phenotype (Schiaffino & Reggiani, 2011), therefore, opening possibilities for exploiting in vitro experimentation to unveil the mechanisms behind developmental establishment of this relationship and possible therapies for specific muscle type atrophies seen in sarcopenia, muscle disuse and congenital myopathies.

When evaluating the impact of electrical stimulation on myotube maturation, the frequency and duration of stimulation can be modified. In one study, regardless of the frequency, the developed 3D human skeletal muscle model showed an amelioration of structural distribution of dystrophin in multinucleated myotubes and increase in cross-sectional area and force production. The same study revealed that increasing frequency of electrical stimulation from 1 to 10 Hz induced an increase in hypertrophic phenotype of artificial skeletal muscle and upregulation of MTORC1, a major regulator of muscle mass in vivo (Khodabukus et al., 2019). Overall positive effect of electrical stimulation on skeletal muscle culture in vitro is demonstrated by a number of studies where in response to electrical stimulation, the myoblasts exhibit enhanced differentiation demonstrated by increase in the expression of MyoD and Desmin genes (Flaibani et al., 2009; Serena et al., 2008; Zhuang et al., 2020). Nonetheless, electrical stimulation alone does not adequately reflect naturally occurring communication between MNs and myofibers due to the lack of contribution of neuron-derived trophic factors and NMJ formation and stabilization.

5. Conclusions and future perspectives

The importance of having access to valid in vitro human skeletal muscle models has been driving the skeletal muscle modeling research of many groups around the world. The advances in bioengineering, stem cell and biomaterial technologies had recently allowed for development of increasingly more complex muscle constructs that can better mimic the native skeletal muscle tissue. With much progress at hand, there are still challenges in the field that require closer look at several aspects.

For instance, current in vitro approaches vary greatly between research groups in terms of cellular composition where a variety of human cells (primary or myogenic cell line) are co-cultured with murine counterparts of either cellular or organotypic nature. This can lead to variable experimental conclusions that can be drawn from different in vitro constructs depending on the experimental set up. A more careful standardization of culture conditions can be a valuable step in acquiring a more relevant data for human-specific studies.

To counter this problem, recent progress in the field of pluripotent stem cells have enabled the research groups to use either hESC or hiPSC in their experimental approach. The availability of myogenic hiPSC differentiation have become extremely useful due to the possibility of studying diseased muscle constructs using patient-derived hiPSCs. On the same line, hiPSC technology has also opened the possibility to derive multiple lineages of human cells eventually with isogenic origin, including MNs. The breakthrough in studying human NMJ has been made possible with the ability to derive MNs from hiPSCs. Until recently, the NMJ modeling relied on the introduction of neural counterpart derived from animal models; however, the availability of hiPSC-derived MNs can further our knowledge of human NMJ formation during development and expand the insights into pathophysiological interactions between different cell types. Additional benefit of co-culture of neural and myogenic counterparts is related to the recapitulation of physiological interaction necessary for enhanced contractile activity and maturation of the myogenic tissue.

Due the complexity of system, current methods are aimed at increased control of 3D construct assembly. Specifically, 3D bioprinting has offered spaciotemporal control of both the environment and cell type distribution within 3D in vitro artificial muscle. This has important implications due to structural and cellular complexity of the myofibers and their interaction with the other cellular components within the instructive ECM. We expect that in the future more effort will be deployed into development of approaches allowing control of cell assembly and cell-cell interaction for better mimic the native tissue.

In this chapter we focused on the advanced development of in vitro models of skeletal muscle. These models can be used for drug screening, for disease modeling, for investigation of electrophysiological or mechanical properties of the muscle, for studying interaction of myofibers with other cellular components, as well as to mimic in vitro skeletal muscle regeneration. However, many of the described in vitro models have the potential to be applied in translational medicine approaches. Indeed, some models, enable the reconstruction of bulk artificial muscle that can be transplanted in vivo. A variety of biocompatible, biodegradable biomaterials are available for use as acellular or recellularized scaffolds which can deliver not only bioactive components necessary for tissue regeneration but also be an active site of regeneration and tissue formation. This approach can be of great value for volumetric muscle loss traumas where an extensive damage within the muscle exceeds the capacity of skeletal muscle tissue regenerative machinery. However, current in vitro bulk artificial muscle production is limiting the size of the fabricated tissue constructs, and great effort is growing in the scientific community to overcome such limitation. For successful VML transplantation studies the diffusion limit has to be considered with techniques aiming at decrease of necrotic core formation leading the development of human-size relative constructs in the near future.

Acknowledgments

This work was supported by IRP Consolidator Grant 2021 to A.U. (21/05Irp), AFM-Telethon Research Project 2020 (23284), LifeLab Program 'Consorzio per la Ricerca Sanitaria' (CORIS) of the Veneto Region, GOSH Children's Charity grant VS0419 and the NIHR GOSH BRC to N.E. We are grateful to Dr. Francesca Cecchinato and Beatrice Auletta for their time and creativity in helping withthe table and the figure.

References

Ahmad, K., Shaikh, S., Ahmad, S. S., Lee, E. J., & Choi, I. (2020). Cross-talk between extracellular matrix and skeletal muscle: Implications for myopathies. *Frontiers in Pharmacology*, *11*(February), 1–8. https://doi.org/10.3389/fphar.2020.00142.

Alhindi, A., et al. (2021). Terminal Schwann cells at the human neuromuscular junction. *Brain Communications*, *3*(2), 1–12. https://doi.org/10.1093/braincomms/fcab081.

Allen, R. E., Sheehan, S. M., Taylor, R. G., Kendall, T. L., & Rice, G. M. (1995). Hepatocyte growth factor activates quiescent skeletal muscle satellite cells in vitro. *Journal of Cellular Physiology*, *165*(2), 307–312. https://doi.org/10.1002/jcp.1041650211.

Alvarez-Suarez, P., Gawor, M., & Prószyński, T. J. (2020). Perisynaptic Schwann cells—The multitasking cells at the developing neuromuscular junctions. *Seminars in Cell and Developmental Biology*, *104*, 31–38. https://doi.org/10.1016/j.semcdb.2020.02.011.

Amthor, H., Christ, B., Weil, M., & Patel, K. (1998). The importance of timing differentiation during limb muscle development. *Current Biology*, *8*(11), 642–652. https://doi.org/10.1016/S0960-9822(98)70251-9.

Arnold, L., Henry, A., Poron, F., Baba-Amer, Y., van Rooijen, N., Plonquet, A., et al. (2007). Inflammatory monocytes recruited after skeletal muscle injury switch into antiinflammatory macrophages to support myogenesis. *Journal of Experimental Medicine*, *204*(5), 1057–1069.

Baghdadi, M. B., Firmino, J., Soni, K., Evano, B., Di Girolamo, D., Mourikis, P., et al. (2018). Notch-induced MiR-708 antagonizes satellite cell migration and maintains quiescence. *Cell Stem Cell*, *23*(6), 859–868.e5. https://doi.org/10.1016/j.stem.2018.09.017.

Baiguera, S., Del Gaudio, C., Di Nardo, P., Manzari, V., Carotenuto, F., & Teodori, L. (2020). 3D printing decellularized extracellular matrix to design biomimetic scaffolds for skeletal muscle tissue engineering. *BioMed Research International*, *2020*. https://doi.org/10.1155/2020/2689701.

Bakooshli, M. A., Lippmann, E. S., Mulcahy, B., Iyer, N., Nguyen, C. T., Tung, K., et al. (2019). A 3D culture model of innervated human skeletal muscle enables studies of the adult neuromuscular junction. *eLife*, *8*, 1–29. https://doi.org/10.7554/eLife.44530.

Bentzinger, C. F., Wang, Y. X., von Maltzahn, J., Soleimani, V. D., Yin, H., & Rudnicki, M. A. (2013). Fibronectin regulates Wnt7a signaling and satellite cell expansion. *Cell Stem Cell*, *12*(1), 75–87. https://doi.org/10.1016/j.stem.2012.09.015.

Besse, L., Sheeba, C. J., Holt, M., Labuhn, M., Wilde, S., Feneck, E., et al. (2020). Individual limb muscle bundles are formed through progressive steps orchestrated by adjacent connective tissue cells during primary myogenesis. *Cell Reports*, *30*(10), 3552–3565. e6. https://doi.org/10.1016/j.celrep.2020.02.037.

Biressi, S, et al. (2007). Cellular heterogeneity during vertebrate skeletal muscle development. *Developmental Biology, 308*(2), 281–293. https://doi.org/10.1016/j.ydbio.2007.06.006.

Bonaldo, P., & Sandri, M. (2013). Cellular and molecular mechanisms of muscle atrophy. *Disease Models & Mechanisms, 6*(1), 25–39. https://doi.org/10.1242/dmm.010389.

Boso, D., Carraro, E., Maghin, E., Todros, S., Dedja, A., Giomo, M., et al. (2021). Porcine decellularized diaphragm hydrogel: A new option for skeletal muscle malformations. *Biomedicine, 9*(7), 1–18. https://doi.org/10.3390/biomedicines9070709.

Boso, D., Maghin, E., Carraro, E., Giagante, M., Pavan, P., & Piccoli, M. (2020). Extracellular matrix-derived hydrogels as biomaterial for different skeletal muscle tissue replacements. *Materials, 13*(11), 2483. https://doi.org/10.3390/ma13112483.

Brack, A. S., Conboy, I. M., Conboy, M. J., Shen, J., & Rando, T. A. (2008). A temporal switch from notch to Wnt signaling in muscle stem cells is necessary for normal adult myogenesis. *Cell Stem Cell, 2*(1), 50–59. https://doi.org/10.1016/j.stem.2007.10.006.

Buono, R., Vantaggiato, C., Pisa, V., Azzoni, E., Bassi, M. T., Brunelli, S., et al. (2012). Nitric oxide sustains long-term skeletal muscle regeneration by regulating fate of satellite cells via signaling pathways requiring Vangl2 and cyclic GMP. *Stem Cells, 30*(2), 197–209. https://doi.org/10.1002/stem.783.

Bushby, K. M. D., Collins, J., & Hicks, D. (2014). Collagen type VI myopathies. *Advances in Experimental Medicine and Biology, 802*, 185–199. https://doi.org/10.1007/978-94-7-7893-1_12.

Cescon, M., et al. (2018). Collagen VI is required for the structural and functional integrity of the neuromuscular junction. *Acta Neuropathologica, 136*(3), 483–499. https://doi.org/10.1007/s00401-018-1860-9.

Cetin, H., Beeson, D., Vincent, A., & Webster, R. (2020). The structure, function, and physiology of the fetal and adult acetylcholine receptor in muscle. *Frontiers in Molecular Neuroscience, 13*(September), 1–14. https://doi.org/10.3389/fnmol.2020.581097.

Chal, J., & Pourquié, O. (2017). Making muscle: Skeletal myogenesis in vivo and in vitro. *Development (Cambridge), 144*(12), 2104–2122. https://doi.org/10.1242/dev.151035.

Chapman, M. A., Meza, R., & Lieber, R. L. (2016). Skeletal muscle fibroblasts in health and disease. *Differentiation, 92*(3), 108–115. https://doi.org/10.1016/j.diff.2016.05.007.

Cheema, U., Brown, R., Vivek Mudera, Y., Shi, Y., Mcgrouther, G., & Goldspink, G. (2005). Mechanical signals and IGF-I gene splicing in vitro in relation to development of skeletal muscle. *Journal of Cellular Physiology, 202*(1), 67–75. https://doi.org/10.1002/jcp.20107.

Chen, X., Wenqiang, D., Cai, Z., Ji, S., Dwivedi, M., Chen, J., et al. (2020). Uniaxial stretching of cell-laden microfibers for promoting C2C12 myoblasts alignment and myofibers formation. *ACS Applied Materials and Interfaces, 12*(2), 2162–2170. https://doi.org/10.1021/acsami.9b22103.

Cimetta, E., Pizzato, S., Bollini, S., Serena, E., De Coppi, P., & Elvassore, N. (2009). Production of arrays of cardiac and skeletal muscle myofibers by micropatterning techniques on a soft substrate. *Biomedical Microdevices, 11*(2), 389–400. https://doi.org/10.1007/s10544-008-9245-9.

Conboy, I. M., & Rando, T. A. (2002). The regulation of notch signaling controls satellite cell activation and cell fate determination in postnatal myogenesis. *Developmental Cell, 3*(3), 397–409. https://doi.org/10.1016/S1534-5807(02)00254-X.

Cornelison, D. D. W., Wilcox-Adelman, S. A., Goetinck, P. F., Rauvala, H., Rapraeger, A. C., & Olwin, B. B. (2004). Essential and separable roles for Syndecan-3 and Syndecan-4 in skeletal muscle development and regeneration. *Genes and Development, 18*(18), 2231–2236. https://doi.org/10.1101/gad.1214204.

Costamagna, D., Mommaerts, H., Sampaolesi, M., & Tylzanowski, P. (2016). Noggin inactivation affects the number and differentiation potential of muscle progenitor cells in vivo. *Scientific Reports, 6*(February), 1–16. https://doi.org/10.1038/srep31949.

Costantini, M., Testa, S., Mozetic, P., Barbetta, A., Fuoco, C., Fornetti, E., et al. (2017). Microfluidic-enhanced 3D bioprinting of aligned myoblast-laden hydrogels leads to functionally organized myofibers in vitro and in vivo. *Biomaterials, 131*, 98–110.

Csapo, R., Gumpenberger, M., & Wessner, B. (2020). Skeletal muscle extracellular matrix— What do we know about its composition, regulation, and physiological roles? A narrative review. *Frontiers in Physiology, 11*(March), 1–15. https://doi.org/10.3389/fphys.2020.00253.

Denes, L. T., Riley, L. A., Mijares, J. R., Arboleda, J. D., McKee, K., Esser, K. A., et al. (2019). Culturing C2C12 myotubes on micromolded gelatin hydrogels accelerates myotube maturation. *Skeletal Muscle, 9*(1), 1–10. https://doi.org/10.1186/s13395-019-0203-4.

Deries, M., & Thorsteinsdóttir, S. (2016). Axial and limb muscle development: Dialogue with the neighbourhood. *Cellular and Molecular Life Sciences, 73*(23), 4415–4431. https://doi.org/10.1007/s00018-016-2298-7.

Duxson, M. J., Usson, Y., & Harris, A. J. (1989). The origin of secondary myotubes in mammalian skeletal muscles: Ultrastructural studies. *Development, 107*(4), 743–750. https://doi.org/10.1242/dev.107.4.743.

Engler, A. J., Griffin, M. A., Sen, S., Bönnemann, C. G., Lee Sweeney, H., & Discher, D. E. (2004). Myotubes differentiate optimally on substrates with tissue-like stiffness: Pathological implications for soft or stiff microenvironments. *Journal of Cell Biology, 166*(6), 877–887. https://doi.org/10.1083/jcb.200405004.

Exeter, D., & Connell, D. A. (2010). Skeletal muscle: Functional anatomy and pathophysiology. *Seminars in Musculoskeletal Radiology, 14*(2), 97–105. https://doi.org/10.1055/s-0030-1253154.

Flaibani, M., Boldrin, L., Cimetta, E., Piccoli, M., De Coppi, P., & Elvassore, N. (2009). Muscle differentiation and myotubes alignment is influenced by micropatterned surfaces and exogenous electrical stimulation. *Tissue Engineering Part A, 15*(9), 2447–2457. https://doi.org/10.1089/ten.tea.2008.0301.

Forcina, L., Cosentino, M., & Musarò, A. (2020). Mechanisms regulating muscle regeneration: Insights into the interrelated and time-dependent phases of tissue healing. *Cell, 9*(5). https://doi.org/10.3390/cells9051297.

Garg, K., & Boppart, M. D. (2016). Influence of exercise and aging on extracellular matrix composition in the skeletal muscle stem cell niche. *Journal of Applied Physiology, 121*(5), 1053–1058. https://doi.org/10.1152/japplphysiol.00594.2016.

Garg, K., Corona, B. T., & Walters, T. J. (2015). Therapeutic strategies for preventing skeletal muscle fibrosis after injury. *Frontiers in Pharmacology, 6*(87), 1–9. https://doi.org/10.3389/fphar.2015.00087.

Gattazzo, F., Urciuolo, A., & Bonaldo, P. (2014). Extracellular matrix: A dynamic microenvironment for stem cell niche. *Biochimica et Biophysica Acta - General Subjects, 1840*(8), 2506–2519. https://doi.org/10.1016/j.bbagen.2014.01.010.

Gilbert, P. M., Havenstrite, K. L., Magnusson, K. E. G., Sacco, A., Leonardi, N. A., Kraft, P., et al. (2010). Substrate elasticity regulates skeletal muscle stem cell self-renewal in culture. *Science, 329*(5995), 1078–1081. https://doi.org/10.1126/science.1191035.

Gilbert-Honick, J., Iyer, S. R., Somers, S. M., Takasuka, H., Lovering, R. M., Wagner, K. R., et al. (2020). Engineering 3D skeletal muscle primed for neuromuscular regeneration following volumetric muscle loss. *Biomaterials, 255*(June), 120154. https://doi.org/10.1016/j.biomaterials.2020.120154.

Gillies, A. R., Chapman, M. A., Bushong, E. A., Deerinck, T. J., Ellisman, M. H., & Lieber, R. L. (2017). High resolution three-dimensional reconstruction of fibrotic skeletal muscle extracellular matrix. *Journal of Physiology, 595*(4), 1159–1171. https://doi.org/10.1113/JP273376.

Giobbe, G. G., Crowley, C., Luni, C., Campinoti, S., Khedr, M., Kretzschmar, K., et al. (2019). Extracellular matrix hydrogel derived from decellularized tissues enables endodermal organoid culture. *Nature Communications*, *10*(1). https://doi.org/10.1038/s41467-019-13605-4.

Gresham, R. C. H., Bahney, C. S., & Kent Leach, J. (2021). Growth factor delivery using extracellular matrix-mimicking substrates for musculoskeletal tissue engineering and repair. *Bioactive Materials*, *6*(7), 1945–1956. https://doi.org/10.1016/j.bioactmat.2020.12.012.

Gromova, A., & La Spada, A. R. (2020). Harmony lost: Cell–cell communication at the neuromuscular junction in motor neuron disease. *Trends in Neurosciences*, *43*(9), 709–724. https://doi.org/10.1016/j.tins.2020.07.002.

Guimarães, C., Gasperini, L., Marques, A., & Reis, R. (2020). The stiffness of living tissues and its implications for tissue engineering. *Nature Reviews Materials*, (5), 351–370. https://doi.org/10.1038/s41578-019-0169-1. In press.

Guiraud, S., Migeon, T., Ferry, A., Chen, Z., Ouchelouche, S., Verpont, M. C., et al. (2017). HANAC Col4a1 mutation in mice leads to skeletal muscle alterations due to a primary vascular defect. *American Journal of Pathology*, *187*(3), 505–516. https://doi.org/10.1016/j.ajpath.2016.10.020.

Halper, J., & Kjaer, M. (2014). Basic components of connective tissues and extracellular matrix: Elastin, fibrillin, fibulins, fibrinogen, fibronectin, laminin, tenascins and thrombospondins. *Advances in Experimental Medicine and Biology*, *802*(January 2014), 31–47. https://doi.org/10.1007/978-94-7-7893-1_3.

Heher, P., Maleiner, B., Prüller, J., Teuschl, A. H., Kollmitzer, J., Monforte, X., et al. (2015). A novel bioreactor for the generation of highly aligned 3D skeletal muscle-like constructs through orientation of fibrin via application of static strain. *Acta Biomaterialia*, *24*(September), 251–265. https://doi.org/10.1016/j.actbio.2015.06.033.

Heikkinen, A., Härönen, H., Norman, O., & Pihlajaniemi, T. (2020). Collagen XIII and other ECM components in the assembly and disease of the neuromuscular junction. *Anatomical Record*, *303*(6), 1653–1663. https://doi.org/10.1002/ar.24092.

Hu, X., Charles, J. P., Akay, T., Hutchinson, J. R., & Blemker, S. S. (2017). Are mice good models for human neuromuscular disease? Comparing muscle excursions in walking between mice and humans. *Skeletal Muscle*, *7*(1), 1–14. https://doi.org/10.1186/s13395-017-0143-9.

Huey, K. A. (2018). Potential roles of vascular endothelial growth factor during skeletal muscle hypertrophy. *Exercise and Sport Sciences Reviews*, *46*(3), 195–202. https://doi.org/10.1249/JES.0000000000000152.

Hurren, B, et al. (2015). First neuromuscular contact correlates with onset of primary myogenesis in rat and mouse limb muscles. *PLoS ONE*, *10*(7), 1–18. https://doi.org/10.1371/journal.pone.0133811.

Jana, S., Lan, S. K., & Levengood, and Miqin Zhang. (2016). Anisotropic materials for skeletal-muscle-tissue engineering. *Advanced Materials*, *28*(48), 10588–10612. https://doi.org/10.1002/adma.201600240.

Kang, H.-W., Lee, S. J., Ko, I. K., Kengla, C., Yoo, J. J., & Atala, A. (2016). A 3D bioprinting system to produce human-scale tissue constructs with structural integrity. *Nature Biotechnology*, *34*(3), 312–319. https://doi.org/10.1038/nbt.3413.

Kasper, A. M., Turner, D. C., Martin, N. R. W., & Sharples, A. P. (2018). Mimicking exercise in three-dimensional bioengineered skeletal muscle to investigate cellular and molecular mechanisms of physiological adaptation. *Journal of Cellular Physiology*, *233*(3), 1985–1998. https://doi.org/10.1002/jcp.25840.

Kaur, G., & Dufour, J. M. (2012). Cell lines: Valuable tools or useless artifacts. *Spermatogenesis*, *2*(1), 1–5. https://doi.org/10.4161/spmg.19885.

Khodabukus, A., Madden, L., Prabhu, N. K., Koves, T. R., Jackman, C. P., Muoio, D. M., et al. (2019). Electrical stimulation increases hypertrophy and metabolic flux in tissue-engineered human skeletal muscle. *Biomaterials*, *198*(3), 259–269. https://doi. org/10.1016/j.biomaterials.2018.08.058.

Kim, J. H., Seol, Y. J., Ko, I. K., Kang, H. W., Lee, Y. K., Yoo, J. J., et al. (2018). 3D bio-printed human skeletal muscle constructs for muscle function restoration. *Scientific Reports*, *8*(1), 1–15. https://doi.org/10.1038/s41598-018-29968-5.

Latroche, C., Gitiaux, C., Chrétien, F., Desguerre, I., Mounie, R., & Chazaud, B. (2015). Skeletal muscle microvasculature: A highly dynamic lifeline. *Physiology*, *30*(6), 417–427. https://doi.org/10.1152/physiol.00026.2015.

Latroche, C., Weiss-Gayet, M., Muller, L., Gitiaux, C., Leblanc, P., Liot, S., et al. (2017). Coupling between myogenesis and angiogenesis during skeletal muscle regeneration is stimulated by restorative macrophages. *Stem Cell Reports*, *9*(6), 2018–2033. https:// doi.org/10.1016/j.stemcr.2017.10.027.

Li, E. W., McKee-Muir, O. C., & Gilbert, P. M. (2018). *Cellular biomechanics in skeletal muscle regeneration. Current topics in developmental biology. Vol. 126* (1st ed.). Elsevier Inc. https:// doi.org/10.1016/bs.ctdb.2017.08.007.

Maffioletti, S. M., Sarcar, S., Henderson, A. B. H., Mannhardt, I., Pinton, L., Moyle, L. A., et al. (2018). Three-dimensional human IPSC-derived artificial skeletal muscles model muscular dystrophies and enable multilineage tissue engineering. *Cell Reports*, *23*(3), 899–908. https://doi.org/10.1016/j.celrep.2018.03.091.

Martins, F., Miguel, J., Fischer, C., Urzi, A., Vidal, R., Kunz, S., et al. (2020). Self-organizing 3D human trunk neuromuscular organoids. *Cell Stem Cell*, *26*(2), 172–186.e6. https://doi.org/10.1016/j.stem.2019.12.007.

Mathew, S. J., Hansen, J. M., Merrell, A. J., Murphy, M. M., Lawson, J. A., Hutcheson, D. A., et al. (2011). Connective tissue fibroblasts and Tcf4 regulate myogenesis. *Development*, *138*(2), 371–384. https://doi.org/10.1242/dev.057463.

McCrary, Michaela W., Deanna Bousalis, Sahba Mobini, Young Hye Song, and Christine E. Schmidt. 2020. "Decellularized tissues as platforms for in vitro modeling of healthy and diseased tissues." Acta Biomaterialia 111: 1–19. https://doi.org/10.1016/j.actbio. 2020.05.031.

Messina, G., Biressi, S., Monteverde, S., Magli, A., Cassano, M., & Perani, L., et al. (2010). Nfix regulates fetal-specific transcription in developing skeletal muscle. Cell, 140(4): 554–566. https://doi.org/10.1016/j.cell.2010.01.027.

Messina, S., & Sframeli, M. (2020). New treatments in spinal muscular atrophy: Positive results and new challenges. *Journal of Clinical Medicine*, *9*(7), 2222. https://doi.org/ 10.3390/jcm9072222.

Michielin, F., Serena, E., Pavan, P., & Elvassore, N. (2015). Microfluidic-assisted cyclic mechanical stimulation affects cellular membrane integrity in a human muscular dystrophy in vitro model. *RSC Advances*, *5*(119), 98429–98439. https://doi.org/10.1039/c5ra16957g.

Monge, C., DiStasio, N., Rossi, T., Sébastien, M., Sakai, H., Kalman, B., et al. (2017). Quiescence of human muscle stem cells is favored by culture on natural biopolymeric films. *Stem Cell Research & Therapy*, *8*(1), 1–13. https://doi.org/10.1186/s13287-017-0556-8.

Mukund, K., & Subramaniam, S. (2020). Skeletal muscle: A review of molecular structure and function, in health and disease. *Wiley Interdisciplinary Reviews: Systems Biology and Medicine*, *12*(1), 1–46. https://doi.org/10.1002/wsbm.1462.

Murphy, M. M., Lawson, J. A., Mathew, S. J., Hutcheson, D. A., & Kardon, G. (2011). Satellite cells, connective tissue fibroblasts and their interactions are crucial for muscle regeneration. *Development*, *138*(17), 3625–3637. https://doi.org/10.1242/dev.064162.

Narayanan, N., Jiang, C., Wang, C., Uzunalli, G., Whittern, N., Chen, D., et al. (2020). Harnessing Fiber diameter-dependent effects of myoblasts toward biomimetic scaffold-based skeletal muscle regeneration. *Frontiers in Bioengineering and Biotechnology*, *8*(March), 1–12. https://doi.org/10.3389/fbioe.2020.00203.

Neuhaus, P., Oustanina, S., Loch, T., Krüger, M., Bober, E., Dono, R., et al. (2003). Reduced mobility of fibroblast growth factor (FGF)-deficient myoblasts might contribute to dystrophic changes in the musculature of FGF2/FGF6/mdx triple-mutant mice. *Molecular and Cellular Biology*, *23*(17), 6037–6048. https://doi.org/10.1128/mcb.23.17. 6037-6048.2003.

Nguyen, K. T., & West, J. L. (2002). Photopolymerizable hydrogels for tissue engineering applications. *Biomaterials*, *23*(22), 4307–4314. https://doi.org/10.1016/S0142-9612(02) 00175-8.

Osaki, T., Uzel, S. G. M., & Kamm, R. D. (2018). Microphysiological 3D model of amyotrophic lateral sclerosis (ALS) from human IPS-derived muscle cells and optogenetic motor neurons. *Science Advances*, *4*(10), 1–15. https://doi.org/10.1126/ sciadv.aat5847.

Ostrovidov, S., Salehi, S., Costantini, M., Suthiwanich, K., Ebrahimi, M., Sadeghian, R. B., et al. (2019). 3D bioprinting in skeletal muscle tissue engineering. *Small*, *15*(24), 1805530. https://doi.org/10.1002/smll.201805530.

Pan, C., Kumar, C., Bohl, S., Klingmueller, U., & Mann, M. (2009). Comparative proteomic phenotyping of cell lines and primary cells to assess preservation of cell type-specific functions. *Molecular and Cellular Proteomics*, *8*(3), 443–450. https://doi.org/10.1074/ mcp.M800258-MCP200.

Philippos, M., Sambasivan, R., Castel, D., Rocheteau, P., Bizzarro, V., & Tajbakhsh, S. (2012). A critical requirement for notch signaling in maintenance of the quiescent skeletal muscle stem cell state. *Stem Cells*, *30*(2), 243–252. https://doi.org/10. 1002/stem.775.

Powell, C. A., Smiley, B. L., Mills, J., & Vandenburgh, H. H. (2002). Mechanical stimulation improves tissue-engineered human skeletal muscle. *American Journal of Physiology - Cell Physiology*, *283*(5 52–5), 1557–1565. https://doi.org/10.1152/ajpcell.00595.2001.

Quarta, M., Brett, J. O., DiMarco, R., De Morree, A., Boutet, S. C., Chacon, R., et al. (2016). A bioengineered niche preserves the quiescence of muscle stem cells and enhances their therapeutic efficacy HHS public access author manuscript. *Nature Biotechnology*, *34*(7), 752–759. https://doi.org/10.1038/nbt.3576.A.

Raffa, P., Scattolini, V., Gerli, M. F. M., Perin, S., Cui, M., De Coppi, P., et al. (2020). Decellularized skeletal muscles display neurotrophic effects in three-dimensional organotypic cultures. *Stem Cells Translational Medicine*, *9*(10), 1233–1243. https://doi. org/10.1002/sctm.20-0090.

Rigamonti, E., Touvier, T., Clementi, E., Manfredi, A. A., Brunelli, S., & Rovere-Querini, P. (2013). Requirement of inducible nitric oxide synthase for skeletal muscle regeneration after acute damage. *The Journal of Immunology*, *190*(4), 1767–1777. https://doi. org/10.4049/jimmunol.1202903.

van Ry, P. M., Tatiana, M., Fontelonga, P. B.-F., Sarathy, A., Nunes, A. M., & Burkin, D. J. (2017). ECM-related myopathies and muscular dystrophies: Pros and cons of protein therapies. *Comprehensive Physiology*, *7*(4), 1519–1536. https://doi.org/10.1002/cphy.c150033.

Sambasivan, R., Yao, R., Kissenpfennig, A., van Wittenberghe, L., Paldi, A., Gayraud-Morel, B., et al. (2011). Pax7-expressing satellite cells are indispensable for adult skeletal muscle regeneration. *Development*. https://doi.org/10.1242/dev.067587.

Schiaffino, S., & Reggiani, C. (2011). Fiber types in mammalian skeletal muscles. *Physiological Reviews*, *91*(4), 1447–1531. https://doi.org/10.1152/physrev.00031.2010.

Serena, E., Flaibani, M., Carnio, S., Boldrin, L., Vitiello, L., De Coppi, P., et al. (2008). Electrophysiologic stimulation improves myogenic potential of muscle precursor cells grown in a 3D collagen scaffold. *Neurological Research*, *30*(2), 207–214. https://doi. org/10.1179/174313208X281109.

Serena, E., Zatti, S., Reghelin, E., Pasut, A., Cimetta, E., & Elvassore, N. (2010). Soft substrates drive optimal differentiation of human healthy and dystrophic myotubes. *Integrative Biology*, *2*(4), 193. https://doi.org/10.1039/b921401a.

Serena, E., Zatti, S., Zoso, A., Verso, F. L., Saverio Tedesco, F., Cossu, G., et al. (2016). Skeletal muscle differentiation on a chip shows human donor Mesoangioblasts' efficiency in restoring dystrophin in a Duchenne muscular dystrophy model. *Stem Cells Translational Medicine, 5*(12), 1676–1683. https://doi.org/10.5966/sctm.2015-0053.

Slater, C. R. (2017). The structure of human neuromuscular junctions: Some unanswered molecular questions. *International Journal of Molecular Sciences, 18*(10). https://doi.org/10.3390/ijms18102183.

Slater, C. R., & Schiaffino, S. (2008). Innervation of regenerating muscle. *Skeletal Muscle Repair and Regeneration,* 303–334. https://doi.org/10.1007/978-1-4020-6768-6_14.

Smith, I. W., Mikesh, M., Lee, Y. I., & Thompson, W. J. (2013). Terminal Schwann cells participate in the competition underlying neuromuscular synapse elimination. *Journal of Neuroscience, 33*(45), 17724–17736. https://doi.org/10.1523/JNEUROSCI.3339-13.2013.

Sousa-Victor, P., García-Prat, L., & Muñoz-Cánoves, P. (2021). Control of satellite cell function in muscle regeneration and its disruption in ageing. *Nature Reviews Molecular Cell Biology.*

Stearns-Reider, K. M., D'Amore, A., Beezhold, K., Rothrauff, B., Cavalli, L., Wagner, W. R., et al. (2017). Aging of the skeletal muscle extracellular matrix drives a stem cell fibrogenic conversion. *Aging Cell, 16*(3), 518–528. https://doi.org/10.1111/acel.12578.

Steele-Stallard, H. B., Pinton, L., Sarcar, S., Ozdemir, T., Maffioletti, S. M., Zammit, P. S., et al. (2018). Modeling skeletal muscle laminopathies using human induced pluripotent stem cells carrying pathogenic LMNA mutations. *Frontiers in Physiology, 9*(OCT), 1–19. https://doi.org/10.3389/fphys.2018.01332.

Stifani, N. (2014). Motor neurons and the generation of spinal motor neuron diversity. *Frontiers in Cellular Neuroscience, 8*(OCT), 1–22. https://doi.org/10.3389/fncel.2014.00293.

Tabakov, V. Y., Zinov'eva, O. E., Voskresenskaya, O. N., & Skoblov, M. Y. (2018). Isolation and characterization of human myoblast culture in vitro for technologies of cell and gene therapy of skeletal muscle pathologies. *Bulletin of Experimental Biology and Medicine, 164*(4), 536–542. https://doi.org/10.1007/s10517-018-4028-7.

Takeda, T., Kitagawa, K., & Arai, K. (2020). Phenotypic variability and its pathological basis in amyotrophic lateral sclerosis. *Neuropathology, 40*(1), 40–56. https://doi.org/10.1111/neup.12606.

Tatsumi, R., Anderson, J. E., Nevoret, C. J., Halevy, O., & Allen, R. E. (1998). HGF/SF is present in normal adult skeletal muscle and is capable of activating satellite cells. *Developmental Biology, 194*(1), 114–128. https://doi.org/10.1006/dbio.1997.8803.

Tedesco, F. S., Dellavalle, A., Diaz-manera, J., Messina, G., & Cossu, G. (2010). Regenerative potential of skeletal muscle stem cells. *The Journal of Clinical Investigation, 120*(1). https://doi.org/10.1172/JCI40373.

Tidball, J. G., & Wehling-Henricks, M. (2007). Macrophages promote muscle membrane repair and muscle fibre growth and regeneration during modified muscle loading in mice in vivo. *The Journal of Physiology, 578*(1), 327–336. https://doi.org/10.1113/jphysiol.2006.118265.

Trevisan, C., Fallas, M. E. A., Maghin, E., Franzin, C., Pavan, P., Caccin, P., et al. (2019). Generation of a functioning and self-renewing diaphragmatic muscle construct. *Stem Cells Translational Medicine, 8*(8), 858–869. https://doi.org/10.1002/sctm.18-0206.

Urciuolo, A., & De Coppi, P. (2018). Decellularized tissue for muscle regeneration. *International Journal of Molecular Sciences.* https://doi.org/10.3390/ijms19082392.

Urciuolo, A., Poli, I., Brandolino, L., Raffa, P., Scattolini, V., Laterza, C., et al. (2020). Intravital three-dimensional bioprinting. *Nature Biomedical Engineering, 4*(9), 901–915. https://doi.org/10.1038/s41551-020-0568-z.

Urciuolo, A., Quarta, M., Morbidoni, V., Gattazzo, F., Molon, S., Grumati, P., et al. (2013). Collagen VI regulates satellite cell self-renewal and muscle regeneration. *Nature Communications, 4.* https://doi.org/10.1038/ncomms2964.

Vila, O. F., Yihuai, Q., & Vunjak-Novakovic, G. (2020). In vitro models of neuromuscular junctions and their potential for novel drug discovery and development. *Expert Opinion on Drug Discovery, 15*(3), 307–317. https://doi.org/10.1080/17460441.2020.1700225.

Visscher, D. O., Farré-Guasch, E., Helder, M. N., Gibbs, S., Forouzanfar, T., van Zuijlen, P. P., et al. (2016). Advances in bioprinting technologies for craniofacial reconstruction. *Trends in Biotechnology, 34*(9), 700–710. https://doi.org/10.1016/j.tibtech.2016.04.001.

von Maltzahn, J., Florian Bentzinger, C., & Rudnicki, M. A. (2012). Wnt7a–Fzd7 signalling directly activates the Akt/MTOR anabolic growth pathway in skeletal muscle. *Nature Cell Biology, 14*(2), 186–191. https://doi.org/10.1038/ncb2404.

Wang, X., Rivera-Bolanos, N., Jiang, B., & Ameer, G. A. (2019). Advanced functional biomaterials for stem cell delivery in regenerative engineering and medicine. *Advanced Functional Materials, 29*(23), 1–31. https://doi.org/10.1002/adfm.201809009.

Wilson, D., Breen, L., Lord, J. M., & Sapey, E. (2018). The challenges of muscle biopsy in a community based geriatric population. *BMC Research Notes, 11*(1), 1–6. https://doi.org/10.1186/s13104-018-3947-8.

Witzemann, V. (2006). Development of the neuromuscular junction. *Cell and Tissue Research, 326*(2), 263–271. https://doi.org/10.1007/s00441-006-0237-x.

Wu, H., Xiong, W. C., & Mei, L. (2010). To build a synapse: Signaling pathways in neuromuscular junction assembly. *Development, 137*(7), 1017–1033. https://doi.org/10.1242/dev.038711.

Xue, J., Tong, W., & Xia, Y. (2018). Perspective: Aligned arrays of electrospun nanofibers for directing cell migration. *APL Materials, 6*(12). https://doi.org/10.1063/1.5058083.

Yamada, M., Tatsumi, R., Kikuiri, T., Okamoto, S., Nonoshita, S., Mizunoya, W., et al. (2006). Matrix metalloproteinases are involved in mechanical stretch-induced activation of skeletal muscle satellite cells. *Muscle and Nerve, 34*(3), 313–319. https://doi.org/10.1002/mus.20601.

Yang, X., Arber, S., William, C., Li, L., Tanabe, Y., Jessell, T. M., et al. (2001). Patterning of muscle acetylcholine receptor gene expression in the absence of motor innervation. *Neuron, 30*(2), 399–410. https://doi.org/10.1016/S0896-6273(01)00287-2.

Yang, W., & Hu, P. (2018). Skeletal muscle regeneration is modulated by inflammation. *Journal of Orthopaedic Translation, 13*, 25–32. https://doi.org/10.1016/j.jot.2018.01.002.

Yao, Y. (2017). Laminin: Loss-of-function studies. *Cellular and Molecular Life Sciences, 74*(6), 1095–1115. https://doi.org/10.1007/s00018-016-2381-0.

Yin, H., Price, F., & Rudnicki, M. A. (2013). Satellite cells and the muscle stem cell niche. *Physiological Reviews, 93*(1), 23–67. https://doi.org/10.1152/physrev.00043.2011.

Zahari, N. K., Idrus, R. B. H., & Chowdhury, S. R. (2017). Laminin-coated poly(methyl methacrylate) (PMMA) nanofiber scaffold facilitates the enrichment of skeletal muscle myoblast population. *International Journal of Molecular Sciences, 18*(11). https://doi.org/10.3390/ijms18112242.

Zhang, Y., Lahmann, I., Baum, K., Shimojo, H., Mourikis, P., Wolf, J., et al. (2021). Oscillations of delta-like 1 regulate the balance between differentiation and maintenance of muscle stem cells. *Nature Communications, 12*(1). https://doi.org/10.1038/s41467-021-21631-4.

Zhao, Y., Zeng, H., Nam, J., & Agarwal, S. (2009). Fabrication of skeletal muscle constructs by topographic activation of cell alignment. *Biotechnology and Bioengineering, 102*(2), 624–631. https://doi.org/10.1002/bit.22080.

Zhuang, P., An, J., Chua, C. K., & Tan, L. P. (2020). Bioprinting of 3D in vitro skeletal muscle models: A review. *Materials and Design, 193.* https://doi.org/10.1016/j.matdes.2020.108794, 108794.

Printed and bound by CPI Group (UK) Ltd, Croydon, CR0 4YY

08/05/2025

01864965-0001